# Loop Groups, Discrete Versions of Some Classical Integrable Systems, and Rank 2 Extensions

# Recent Titles in This Series

479 **Percy Deift, Luen-Chau Li, and Carlos Tomei,** Loop groups, discrete versions of some classical integrable systems, and rank 2 extensions, 1992

478 **Henry C. Wente,** Constant mean curvature immersions of Enneper type, 1992

477 **George E. Andrews, Bruce C. Berndt, Lisa Jacobsen, and Robert L. Lamphere,** The continued fractions found in the unorganized portions of Ramanujan's notebooks, 1992

476 **Thomas C. Hales,** The subregular germ of orbital integrals, 1992

475 **Kazuaki Taira,** On the existence of Feller semigroups with boundary conditions, 1992

474 **Francisco González-Acuña and Wilbur C. Whitten,** Imbeddings of three-manifold groups, 1992

473 **Ian Anderson and Gerard Thompson,** The inverse problem of the calculus of variations for ordinary differential equations, 1992

472 **Stephen W. Semmes,** A generalization of riemann mappings and geometric structures on a space of domains in $\mathbf{C}^n$, 1992

471 **Michael L. Mihalik and Steven T. Tschantz,** Semistability of amalgamated products and HNN-extensions, 1992

470 **Daniel K. Nakano,** Projective modules over Lie algebras of Cartan type, 1992

469 **Dennis A. Hejhal,** Eigenvalues of the Laplacian for Hecke triangle groups, 1992

468 **Roger Kraft,** Intersections of thick Cantor sets, 1992

467 **Randolph James Schilling,** Neumann systems for the algebraic AKNS problem, 1992

466 **Shari A. Prevost,** Vertex algebras and integral bases for the enveloping algebras of affine Lie algebras, 1992

465 **Steven Zelditch,** Selberg trace formulae and equidistribution theorems for closed geodesics and Laplace eigenfunctions: finite area surfaces, 1992

464 **John Fay,** Kernel functions, analytic torsion, and moduli spaces, 1992

463 **Bruce Reznick,** Sums of even powers of real linear forms, 1992

462 **Toshiyuki Kobayashi,** Singular unitary representations and discrete series for indefinite Stiefel manifolds $U(p,q;\mathbb{F})/U(p-m,q;\mathbb{F})$, 1992

461 **Andrew Kustin and Bernd Ulrich,** A family of complexes associated to an almost alternating map, with application to residual intersections, 1992

460 **Victor Reiner,** Quotients of coxeter complexes and $P$-partitions, 1992

459 **Jonathan Arazy and Yaakov Friedman,** Contractive projections in $C_p$, 1992

458 **Charles A. Akemann and Joel Anderson,** Lyapunov theorems for operator algebras, 1991

457 **Norihiko Minami,** Multiplicative homology operations and transfer, 1991

456 **Michał Misiurewicz and Zbigniew Nitecki,** Combinatorial patterns for maps of the interval, 1991

455 **Mark G. Davidson, Thomas J. Enright and Ronald J. Stanke,** Differential operators and highest weight representations, 1991

454 **Donald A. Dawson and Edwin A. Perkins,** Historical processes, 1991

453 **Alfred S. Cavaretta, Wolfgang Dahmen, and Charles A. Micchelli,** Stationary subdivision, 1991

452 **Brian S. Thomson,** Derivates of interval functions, 1991

451 **Rolf Schön,** Effective algebraic topology, 1991

450 **Ernst Dieterich,** Solution of a non-domestic tame classification problem from integral representation theory of finite groups ($\Lambda = RC_3, v(3) = 4$), 1991

449 **Michael Slack,** A classification theorem for homotopy commutative H-spaces with finitely generated mod 2 cohomology rings, 1991

448 **Norman Levenberg and Hiroshi Yamaguchi,** The metric induced by the Robin function, 1991

(See the AMS catalog for earlier titles)

# MEMOIRS
of the
American Mathematical Society

Number 479


# Loop Groups, Discrete Versions of Some Classical Integrable Systems, and Rank 2 Extensions


Percy Deift
Luen-Chau Li
Carlos Tomei

November 1992 • Volume 100 • Number 479 (second of 4 numbers) • ISSN 0065-9266

**American Mathematical Society**
Providence, Rhode Island

1991 *Mathematics Subject Classification*.
Primary 34, 70, 15.

**Library of Congress Cataloging-in-Publication Data**

Deift, Percy, 1945–
Loop groups, discrete versions of some classical integrable systems, and rank 2 extensions/Percy A. Deift, Luen-Chua Li, Carlos Tomei.
p. cm. – (Memoirs of the American Mathematical Society, ISSN 0065-9266; no. 479)
Includes bibliographical references.
ISBN 0-8218-2540-2
1. Hamiltonian systems. 2. Loops (Group theory) I. Li, Luen-Chau, 1954– . II. Tomei, Carlos. III. Title. IV. Series.
QA3.A57 no. 479
[QA614.83]
510 s–dc20 92-28571
[514′.74] CIP

## Memoirs of the American Mathematical Society

This journal is devoted entirely to research in pure and applied mathematics.

**Subscription information**. The 1992 subscription begins with Number 459 and consists of six mailings, each containing one or more numbers. Subscription prices for 1992 are $292 list, $234 institutional member. A late charge of 10% of the subscription price will be imposed on orders received from nonmembers after January 1 of the subscription year. Subscribers outside the United States and India must pay a postage surcharge of $25; subscribers in India must pay a postage surcharge of $43. Expedited delivery to destinations in North America $30; elsewhere $82. Each number may be ordered separately; *please specify number* when ordering an individual number. For prices and titles of recently released numbers, see the New Publications sections of the *Notices of the American Mathematical Society*.

**Back number information**. For back issues see the *AMS Catalogue of Publications*.

Subscriptions and orders should be addressed to the American Mathematical Society, P. O. Box 1571, Annex Station, Providence, RI 02901-1571. *All orders must be accompanied by payment*. Other correspondence should be addressed to Box 6248, Providence, RI 02940-6248.



*Memoirs of the American Mathematical Society* is published bimonthly (each volume consisting usually of more than one number) by the American Mathematical Society at 201 Charles Street, Providence, RI 02904-2213. Second-class postage paid at Providence, Rhode Island. Postmaster: Send address changes to Memoirs, American Mathematical Society, P. O. Box 6248, Providence, RI 02940-6248.



Printed in the United States of America.
This volume was printed directly from author-prepared copy.
The paper used in this book is acid-free and falls within the guidelines established to ensure permanence and durability. ♾
10 9 8 7 6 5 4 3 2 1 97 96 95 94 93 92

## Table of Contents

## Abstract

The authors show how to interpret recent results of Moser and Veselov on discrete versions of a class of classical integrable systems, in terms of a loop group framework. In this framework the discrete systems appear as time-one maps of integrable Hamiltonian flows. Earlier results of Moser on isospectral deformations of rank 2 extensions of a fixed matrix, can also be incorporated into their scheme.


---


Key words: Hamiltonian mechanics, integrable systems, loop groups, $R$-matrices, discrete Euler-Arnold equation, billiard map, rank 2 extensions.

## Acknowledgements

The work of the first author was supported in part by NSF Grant DMS-9001857. The work of the second author was supported in part by NSF Grant DMS-8704097. The third author acknowledges the support of $\mathrm{CNP_q}$, Brazil, and the warm hospitality of the Department of Mathematics at Yale University. The first author would also like to acknowledge the support of the Institute for Advanced Study, Princeton, NJ, where part of this work was completed.

## 1. Introduction

In a recent paper, Moser and Veselov [MV] considered a class of discrete systems which arise as the Euler-Lagrange equations for a formal sum

$$\mathcal{S} = \sum_{k \in \mathbb{Z}} L(X_k, X_{k+1}) , \tag{1.1}$$

where the $X_k$ are points on a manifold $M^n$, $L(\cdot,\cdot)$ is a function on $Q^{2n} = M^n \times M^n$, and $k \in \mathbb{Z}$ plays the role of the discrete time. As in the analogous continuum problem

$$\mathcal{S} = \int L(q, \dot{q})\, dt , \tag{1.2}$$

one introduces an associated symplectic structure (see [V], [MV]), and the Euler-Lagrange equations give rise to a mapping

$$\Psi : (X_k, X_{k+1}) \to (X_{k+1}, X_{k+2}) \tag{1.3}$$

which is symplectic with respect to the structure. Appropriate choices of $L$ and $M^n$ lead to a wide variety of dynamical systems with many remarkable properties and also of considerable mathematical and physical interest (see [MV] and the references therein).

Of particular interest in [MV] is the case $M^n = O(N)$, $n = N(N-1)/2$, and

$$L(X, Y) = \operatorname{tr} XJY^T , \tag{1.4}$$

where $J$ is a positive symmetric matrix which may be taken to be diagonal without loss of generality. This problem, introduced in [V], converges in the continuum limit to the force-free motion of a rigid body as generalized by Arnold to arbitrary dimensions. The remarkable discovery in [MV] is that the Euler-Lagrange equations for (1.4) can be solved by a QR-type algorithm. Recall that the classical QR-algorithm for diagonalizing matrices (see, for example, [Wi]) proceeds by factoring a real invertible matrix $M$ (uniquely) into a product $M = QR$ of an orthogonal matrix $Q$ and an upper triangular matrix $R$ with positive diagonal entries. The basic step in the algorithm consists in mapping

**Received by editor January 31, 1991.**

$$M = QR \mapsto M' \equiv RQ \ , \tag{1.5}$$

which is isospectral, as $M' = Q^{-1}MQ$. In the case of (1.4), $M$ is now a particular quadratic matrix polynomial

$$M = M(\lambda) = E_0 + \lambda E_1 + \lambda^2 E_2 \ , \tag{1.6}$$

where the coefficients $E_i$, $i = 0, 1, 2$, depend on $X$ and $Y$ in an explicit way (see below). and the role of the QR factorization is played by a particular, unique factorization of $M$ into first order matrix polynomials

$$M(\lambda) = (B_0 + B_1\lambda)(C_0 + C_1\lambda) \ . \tag{1.7}$$

Exchanging the factors,

$$M'(\lambda) \equiv (C_0 + C_1\lambda)(B_0 + B_1\lambda) \tag{1.8}$$

implements the mapping $\Psi$ associated with (1.4). There are, in addition, further consequences:

(i) The isospectral nature of the map $M(\lambda) \mapsto M'(\lambda)$ implies the 'integrability' of $\Psi$, and

(ii) $\Psi$ linearizes on the Jacobi variety of the associated curve $\{(\lambda, \eta) \in \mathbb{C}^2 : \det(M(\lambda) - \eta) = 0\}$.

Over the last decade, following the seminal work of Symes ([Sy1], [Sy2]), the QR algorithm, and related algorithms such as the LU algorithm and the Cholesky algorithm, have come to be understood (see, for example, [Chu], [Wa], [DLNT], [DLT]) as time-one maps of completely integrable Hamiltonian systems closely related with the Toda flow and its generalizations. The underlying symplectic structures are Lie-Poisson structures on the coadjoint orbits of particular Lie-algebras, which in turn are double Lie-algebras carrying classical R-matices satisfying the (modified) Yang-Baxter equation. Recall that $R$-matrix theory (see, for example, [STS], [FT]; see also [DL] for a more pedestrian account) gives a natural explanation of the existence of many commuting integrals, and also leads in a natural way to an explicit solutions procedure by factorization for a class of invariant Hamiltonian flows, as in (1.5) above. For the convenience of the reader we present a brief summary of the relevant results of classical $R$-matrix theory in the Appendix.

The main task of this paper is to give a Lie-algebraic interpretation of the results in [MV]. The underlying algebra turns out to be a loop algebra with an associated classical $R$-matrix which is an appropriate generalization of the $R$-matrix arising in the dynamical theory of the Cholesky algorithm, as described in [DLT]. The discrete systems of Moser and Veselov are time-one maps of integrable Hamiltonian systems, and the solution procedure (1.7), (1.8), and its analogs for all the systems considered in [MV], is precisely the factorization procedure, here of Riemann-Hilbert type, suggested by the general theory of classical $R$-matrices. A single loop algebra suffices to describe all the systems in [MV]: all that differs from one system to the next, is the particular choice of coadjoint orbit of the associated loop group (however, see §3).

In order to describe our results in greater detail, we need more information from [MV]. In the case (1.4),

$$\mathcal{S} = \sum_k \operatorname{tr} X_k J X_{k+1}^T , \tag{1.9}$$

the Euler-Lagrange equations take the form

$$X_{k+1} J + X_{k-1} J = \Lambda_k X_k \tag{1.10}$$

where $\Lambda_k = \Lambda_k^T$ is a matrix Lagrange multipler: $\Lambda_k$ is uniquely determined by $X_{k-1}$, $X_k$, $X_{k+1}$, but not uniquely determined by $X_{k-1}, X_k$. Thus the discrete Euler-Lagrange equations lead in general to a correspondence $(X_{k-1}, X_k) \mapsto (X_k, X_{k+1})$, and the choice of a particular mapping $\Psi$ is equivalent to the choice of a particular branch of the correspondence. Setting

$$\omega_k = X_k^T X_{k-1} \in O(N) , \tag{1.11}$$

and using $\Lambda_k = \Lambda_k^T$, equation (1.10) can be rewritten as a "discrete Euler-Arnold equation" (see [V]),

$$M_{k+1} = \omega_k M_k \omega_k^T , \tag{1.12}$$

where

$$M_k = \omega_k^T J - J\omega_k \in o(N) . \tag{1.13}$$

In the variables $(\omega_k, M_k)$, the nature of the above correspondence is refelcted in the fact that $\omega_k$ is not uniquely determined by $M_k$ through (1.13). The choice of a particular mapping $\Psi$ reduces, given $M_k$, to a *particular* choice of matrix $\omega_k \in O(N)$ in (1.13).

Moser and Veselov proceed as follows. They consider the closed 2-form

$$\omega = \text{ tr } dXJ \wedge dY^T = \sum_{i,j} J_j \, dX_{ij} \wedge dY_{ij} \tag{1.14}$$

restricted to $Q^{2n} = O(N) \times O(N)$. A straightforward, but somewhat tedious, computation shows that,

$$\begin{gathered} \omega \text{ is nondegenerate at } (X,Y) \in O(N) \times O(N) \\ \Leftrightarrow \\ \lambda + \lambda' \neq 0 \text{ for all } \lambda, \lambda' \in \text{ spec}(Y^T X J^{-1}) . \end{gathered} \tag{1.15}$$

In other words, $\omega$ is nondegenerate at a point $(X_{k-1}, X_k) \in Q^{2n}$ only if $\lambda + \lambda' \neq 0$ for all (generalized) eigenvalues $\lambda, \lambda'$, $\det(\omega_k - \lambda J) = \det(\omega_k - \lambda' J) = 0$, where again $\omega_k = X_k^T X_{k-1}$. The basic observation in [MV] is that (1.13) is equivalent to the matrix polynomial factorization

$$M_k(\lambda) \equiv I - \lambda M_k - \lambda^2 J^2 = (\omega_k^T + \lambda J)(\omega_k - \lambda J) \tag{1.16}$$

and switching factors

$$\begin{aligned} M_{k+1}(\lambda) &\equiv (\omega_k - \lambda J)(\omega_k^T + \lambda J) \\ &= I - \lambda M_{k+1} - \lambda^2 J^2 \end{aligned} \tag{1.17}$$

yields $M_{k+1} = \omega_k M_k \omega_k^T$, by (1.12). For $(X_{k-1}, X_k)$ satisfying (1.15), the above factorization for $M_k(\lambda)$ has the property that for

$$\begin{aligned} S &\equiv \{\lambda : \det(I - \lambda M_k - \lambda^2 J^2) = 0\} , \\ S_+ &\equiv \{\lambda : \det(\omega_k - \lambda J) = 0\} , \\ S_- &\equiv \{\lambda : \det(\omega_k^T + \lambda J) = 0\} , \end{aligned} \tag{1.18}$$

we have

$$S = S_+ \cup S_- , \quad \overline{S}_\pm = S_\pm , \quad S_+ = -S_- \quad \text{and} \quad S_+ \cap S_- = \emptyset . \tag{1.19}$$

Conversely, using a technique which is closely related to the solution of a well-known matrix Riccati equation arising in control theory (see, for example, [S-H], [BG-M]), the authors show that given a quadratic pencil $I - \lambda M - \lambda^2 J^2$, for which the associated spectrum $S$ has a splitting $S_\pm$ satisfying (1.19), then there exists a unique factorization

$$I - \lambda M - \lambda^2 J^2 = (\omega^T + \lambda J)(\omega - \lambda J)$$

with $S_+ = \{\lambda : \det(\omega - \lambda J) = 0\}$ and $S_- = \{\lambda : \det(\omega^T + \lambda J) = 0\}$. This leads the authors to the following procedure for the solution of the Euler-Lagrange equations, and hence to a particular branch $\Psi$ of the correspondence determined by (1.4): given $(X_{-1}, X_0) \in Q^{2n}$ satisfying (1.15), set

$$M_0(\lambda) = I - \lambda M_0 - \lambda^2 J^2 = (\omega_0^T + \lambda J)(\omega_0 - \lambda J) , \quad \omega_0 = X_0^T X_{-1} . \tag{1.20}$$

Exchanging factors

$$M_1(\lambda) = I - \lambda M_1 - \lambda^2 J^2 \equiv (\omega_0 - \lambda J)(\omega J + \lambda J) , \tag{1.21}$$

is an isospectral-action,

$$\det M_1(\lambda) = \det M_0(\lambda)$$

and hence $S(M_1(\lambda)) = S(M_0(\lambda))$ has a splitting

$$S_\pm(M_1(\lambda)) \equiv S_\pm(M_0(\lambda))$$

which clearly satisfies (1.19). It follows that $M_1(\lambda)$ has a (unique) factorization

$$M_1(\lambda) = (\omega_1^T + \lambda J)(\omega_1 - \lambda J) , \quad \omega_1 \in O(N) \tag{1.22}$$

with $S_+(M_1(\lambda)) = \{\lambda: \det(\omega_1 - \lambda J) = 0\}$, $S_-(M_1(\lambda)) = \{\lambda: \det(\omega_1^T + \lambda J) = 0\}$. Then

$$M_2(\lambda) \equiv (\omega_1 - \lambda J)(\omega_1^T + \lambda J) \tag{1.23}$$

etc., factoring $M_k(\lambda)$ at each stage according to the fixed spectral decomposition $S = S_+(M_0(\lambda)) \cup S_-(M_0(\lambda))$.

In the loop group approach, the simplest situation (see Section 2) arises when

$$S_+ \subset \{\lambda: \text{ Re } \lambda > 0\}\ ,\ S_- \subset \{\lambda: \text{ Re } \lambda < 0\}\ . \tag{1.24}$$

In this case one observes that

$$\widetilde{M}(\lambda) = \frac{I - \lambda M - \lambda^2 J^2}{1 - \lambda^2} \tag{1.25}$$

is a loop on the compactification $\dot{\Sigma}$ of $\Sigma = iR$ with values in $G\ell(N, \mathbb{C})$, satisfying the reality condition

$$\widetilde{M}(\bar{\lambda}) = \overline{\widetilde{M}}(\lambda)\ , \tag{1.26}$$

and the asymptotic condition,

$$\widetilde{M}(\infty) \text{ is diagonal and strictly positive.} \tag{1.27}$$

One also observes that

$$\widetilde{M}(\lambda) = \left(\frac{\omega^T + \lambda J}{1 + \lambda}\right)\left(\frac{\omega - \lambda J}{1 - \lambda}\right) \tag{1.28}$$

is the unique factorization of $\widetilde{M}(\lambda)$ into a product of factors which are,

$$\text{analytic and invertible for Re } \lambda \geq 0,\ \text{Re } \lambda \leq 0 \text{ respectively,} \tag{1.29}$$

and satisfy,

$$\left(\frac{\omega^T + \lambda J}{1 + \lambda}\right)\Bigg|_{\lambda = \infty} = \left(\frac{\omega - \lambda J}{1 - \lambda}\right)\Bigg|_{\lambda = \infty} = \sqrt{\widetilde{M}(\infty)}\ . \tag{1.30}$$

Based on these observations we introduce the connected loop group

$$G \equiv \{g : g \text{ is a smooth loop from } \dot{\Sigma} \text{ to } G\ell(N, \mathbb{C}),$$
$$\text{contractible to the identity, and satisfying (1.26) and (1.27)}\}$$

with pointwise multiplication

$$g_1 g_2(\lambda) \equiv g_1(\lambda) g_2(\lambda) , \tag{1.31}$$

and Lie-algebra

$$\underline{g} = \{X : X \text{ is a smooth loop from } \dot{\Sigma} \text{ to } g\ell(N, \mathbb{C}),$$
$$\text{with } X(\infty) \text{ real and diagonal, and satisfying (1.26)}\}$$

with pointwise commutator,

$$[X_1, X_2](\lambda) = [X_1(\lambda), X_2(\lambda)] . \tag{1.32}$$

For $X \in G$, define $\pi_\pm : \underline{g} \to \underline{g}$,

$$\pi_+ X(\lambda) \equiv \lim_{\epsilon \downarrow 0} \lim_{r \to \infty} \int_{ir}^{-ir} X(\lambda') \frac{đ\lambda'}{\lambda' - (\lambda + \epsilon)} , \qquad \lambda \in i\boldsymbol{R} , \tag{1.33}$$

$$\pi_- X(\lambda) \equiv \lim_{\epsilon \downarrow 0} \lim_{r \to \infty} \int_{-ir}^{ir} X(\lambda') \frac{đ\lambda'}{\lambda' - (\lambda - \epsilon)} . \qquad \lambda \in i\boldsymbol{R} . \tag{1.34}$$

(Here and throughout the paper $đ\lambda$ denotes $d\lambda/2\pi i$, etc.) Elements in Ran $\pi_\pm$ have analytic continuations to Re $\lambda > 0$, Re $\lambda < 0$ respectively,

$$\pi_+ + \pi_- = I , \tag{1.35}$$

$$\pi_+ X(\infty) = \pi_- X(\infty) = X(\infty)/2 , \tag{1.36}$$

and

$$R \equiv \pi_+ - \pi_- \tag{1.37}$$

is a classical $R$-matrix on $\underline{g}$ satisfying the modified Yang-Baxter equation. Thus

$$[X_1, X_2]_R(\lambda) \equiv \frac{1}{2}([X_1, RX_2] + [RX_1, X_2])$$

gives a second Lie-bracket on $\underline{g}$, and we denote the associated Lie-algebra and connected Lie-group by $\underline{\tilde{g}}$ and $\widetilde{G}$ respectively. *The Lie-Poisson structure on the coadjoint orbits of* $\widetilde{G}$ *provides the underlying symplectic structure for the problems at hand.*

**Remark 1.38.** Note that $\pi_\pm$ are not projections and Ran $\pi_+ \cap$ Ran $\pi_- \neq \emptyset$. On the other hand Ran $\pi_\pm$ are subalgebras of $\underline{g}$, and hence $G_\pm = e^{\mathrm{Ran}\ \pi_\pm}$ are subgroups of $G$. This situation is similar to the classical Cholesky algorithm (see [DLT]) where one factors a matrix $M \in g\ell(n, \mathbb{C})$,

$$\pi_+ M = \text{strict upper part of } M + \frac{1}{2}\ \mathrm{diag}(M)$$

and

$$\pi_- M = \text{strict lower part of } M + \frac{1}{2}\ \mathrm{diag}(M)\ .$$

We show that

(1) the coadjoint orbit through $\widetilde{M}(\lambda)$, regarded as an element of $\tilde{g}^*$, is finite dimensional and generically of dimension $4N^2 - 4N$. Also, the orbit is naturally isomorphic to a coadjoint orbit of the direct sum of two copies of the semi-direct product of (the identity component) of $G\ell(N, \boldsymbol{R})$ with $g\ell(N, \boldsymbol{R})$.

(2) The Riemann-Hilbert factorization on $\Sigma = i\boldsymbol{R}$

$$e^{t\widetilde{M}_0(\lambda)} = g_+(t, \lambda)\, g_-(t, \lambda)\ , \quad \lambda \in i\boldsymbol{R} \tag{1.38}$$

has a solution for all $t \in \boldsymbol{R}$, where $g_\pm(t, \cdot) \in G_\pm = e^{\mathrm{Ran}\pi_\pm}$ and $\widetilde{M}_0(\lambda) = (I - \lambda M_0 - \lambda^2 J^2)/(1 - \lambda^2)$. By the results of $R$-matrix theory,

$$\widetilde{M}(t, \lambda) \equiv g_+(t, \lambda)^{-1} \widetilde{M}_0(\lambda) g_+(t, \lambda) \tag{1.39}$$

solves the flow

$$\frac{d}{dt}\widetilde{M}(t, \lambda) = \left[(\pi_- \log \widetilde{M}(t, \cdot))(\lambda)\ ,\ \widetilde{M}(t, \lambda)\right]\ , \tag{1.40}$$

and at integer times interpolates the QR-type algorithm of Moser and Veselov,

$$(1 - \lambda^2) M(k, \lambda) = M_k(\lambda)\ , \quad k \in \mathbb{Z}\ , \tag{1.41}$$

(see (1.22) et seq.). Modulo technicalities (see Section 2), (1.40) is a Hamiltonian flow which is *essentially integrable* (see discussion in Section 2(d)) on the (generically) $4N^2 - 4N$ dimensional coadjoint orbit through $\widetilde{M}_0(\lambda)$.

(3) as anticipated from [MV], the flow $t \mapsto \widetilde{M}(t,\lambda) = (I - \lambda M(t) - \lambda^2 J^2)/(1-\lambda^2)$ induces an integrable Hamiltonian flow $t \mapsto M(t)$ on the dual Lie-algebra $o(N)^*$ equipped with the standard Lie-Poisson structure.

In Section 3 we consider the general case where (1.24) may fail and $S = S_+ \cup S_-$ is any decomposition satisfying (1.19). As we show, this leads to a loop algebra, and a Riemann-Hilbert factorization problem, on a skeleton $\Sigma = \bigcup_{j=-q}^{q} \Sigma_j$ composed of lines $\Sigma_j$ (asymptotically) parallel to $iR$ with $\Sigma_0 = iR$ and $\Sigma_j = -\Sigma_{-j}$. All the results of Section 2 extend to this situation, but we focus on the novel aspects of the Riemann-Hilbert problem.

The authors in [MV] also consider the more general problem $S = \Sigma_k \operatorname{tr} X_k J X_{k+1}^T$, where $X_k$ now lies in the Stiefel manifold $V_{m,N}$ of $m \times N$ matrices satisfying $X_k X_k^T = I_m$. This leads to a factorization problem for quadratic matrix polynomials of the form

$$\lambda^2 J^2 + \lambda M_k - X_{k-1}^T X_{k-1} \ . \tag{1.42}$$

In Section 4, we concentrate on the case where $m = 1$, leaving the more general problem, with its many connections to the work of Adams, Harnad and Previato [AHP] and Schilling [S], to a later publication. We focus in particular on the billiard ball problem in an elliptic region $E = \{x : (x, C^{-2}x) \leq 1\}$ in $R^N$, where $C$ is positive and diagonal. Although the variational functional $S = \sum_k |x_k - x_{k-1}|$, $(x_k, C^{-2}x_k) = 1$, is not of the above type, Moser and Veselov show that the solution of the billiard ball problem in $E$ also reduces to a polynomial matrix factorization problem for a quadratic polynomial

$$L(\lambda) = y \otimes y + \lambda x \wedge y - \lambda^2 C^2 \ , \tag{1.43}$$

where

$$(x, C^{-2}x) = 1 \qquad \text{and} \qquad (y,y) = 1 \ . \tag{1.44}$$

Our results in Section 4 are the following.

(1) As before, the coadjoint orbit of $\widetilde{G}$ through $A(\lambda) = L(\lambda)/(1-\lambda^2)$ provides a symplectic structure for the problem and the analog of the factorization (1.38) again leads to

a Hamiltonian flow which interpolates the billiard motion at integer times $t \in 4\mathbb{Z}$. The novel feature here, as in all cases where $m \neq N$, is that the loop $A(\lambda)$ and hence the loop $e^{t \log A(\lambda)}$, is not invertible at $\lambda = 0 \in \Sigma = i\mathbf{R}$. Thus the factorization problem is nonstandard in a fundamentally new way.

(2) The above flow induces a flow on the unit sphere bundle $Y = \{(x, y) : (x, C^{-2}x) = 1, \|y\| = 1\}$ to the ellipse, which is also Hamiltonian in the symplectic structure obtained by restricting the standard two form $\sum_{i=1}^{N} dx_i \wedge dy_i$ on $\mathbf{R}^{2N}$, to the submanifold $Y$. The induced flow is completely integrable.

(3) In the case $N = 2$ the flow takes a particularly simple form,

$$
\begin{aligned}
\frac{d}{dt} y \otimes y &= \beta[x \wedge y \ , \ y \otimes y] \\
\frac{d}{dt} x \wedge y &= \beta[y \otimes y, C^2] \ ,
\end{aligned}
\tag{1.45}
$$

where $\beta$ is a calculable constant. These equations can be reduced further to the pendulum equation and then solved in the standard way in terms of an elliptic function.

In an earlier paper Moser [M] considered a class of integrable systems which give rise to isospectral deformations of rank 2 extensions

$$
M(x, y) = A + a\, x \otimes x + b\, x \otimes y + c\, y \otimes x + d\, y \otimes y \tag{1.46}
$$

of a fixed, real symmetric matrix $A$. Here

$$
\Delta = ad - bc \neq 0 \ . \tag{1.47}
$$

In the final section, Section 5, we show that all the examples considered by Moser can be interpreted in terms of our loop group framework, in the following sense. Associated to every isospectral deformation $t \mapsto M(x(t), y(t))$, there is a loop $A(x, y, \lambda)$, with $(1 - \lambda^2) A(x, y, \lambda)$ quadratic in $\lambda$, and a Hamiltonian flow

$$
\frac{d}{dt} A(x(t), y(t), \lambda) = \left[ (\pi_- F'(A(x(t), y(t), \cdot)))(\lambda) \ , \ A(x(t), y(t), \lambda) \right] \tag{1.48}
$$

arising from a factorization of type (1.38) in which the logarithm is repalced by $F'(\cdot)$, with the property that for the appropriate parameter value $\lambda = \lambda_0$, (1.48) reduces to the given isospectral deformation. In particular we learn that the isospectral deformations considered in [M] can be solved by a Symes-type factorization in an appropriate loop group. Furthermore, the curve $\{(\lambda, \eta) : \det(A(x, y, \lambda) - \eta) = 0\}$ is precisely the curve used by Moser in [M] to linearize the flow.

**Remark.** As noted in the Appendix, the presentation of the basic results of $R$-matrix theory is given with the finite dimensional situation in mind. In the infinite dimensional situation the results remain true after suitable modification. Our point of view is that $R$-matrix theory provides a useful and accurate guide to what is true in the loop group situation, and we provide infinite dimensional versions of the proofs and the definitions as they are needed in the text.

## 2. The discrete Euler-Arnold equation (I)

### (a) Algebras, groups and $R$-matrices

Let $G$, $\underline{g}$, $\pi_+$, and $\pi_-$ be as defined in Section 1. A convenient variable on $\dot{\Sigma}$ is given by

$$z = (\lambda - 1)/(\lambda + 1) \tag{2.1}$$

which takes

$$\left.\begin{aligned} \dot{\Sigma} &\to S^1 = \{|z| = 1\} \\ \{\text{Re } \lambda > 0\} &\to \{|z| < 1\} \\ \{\text{Re } \lambda < 0\} &\to \{|z| > 1\} \\ \infty &\to 1 \\ d\!\!\!{}^-\lambda/(1-\lambda^2) &\to -d\!\!\!{}^- z/2z \\ \text{and has inverse} & \\ \lambda &= (1+z)/(1-z)\ . \end{aligned}\right\} \tag{2.2}$$

Smooth loops on $\dot{\Sigma}$ are mapped to smooth loops on $S^1$,

$$X(z) = \sum_{j=-\infty}^{\infty} X_j z^j \tag{2.3}$$

with $\|X_j\|$ rapidly decreasing as $|j| \to \infty$. We will use both the $\lambda$ variable and the $z$ variable for the loops, depending on convenience. We will, however, consistently abuse notation and denote $X(\lambda(z))$ simply by $X(z)$, etc.

**Notation (2.4)**

We use the standard notation $\displaystyle⨍(\cdot)\frac{d\!\!\!{}^- z}{z-1}$ to denote the Cauchy principal value at $z = 1$,

$$⨍ A(z)\frac{d\!\!\!{}^- z}{z-1} = \lim_{\epsilon \downarrow 0} \int_{|z-1|>\epsilon} A(z)\frac{d\!\!\!{}^- z}{z-1}\ , \tag{2.5}$$

where the latter integral is taken counterclockwise.

Changing variables in (1.34), we have for $z \in S^1$

$$\pi_- X(z) = \lim_{\substack{z'' \to z \\ |z''|>1}} \left[ - \oint X(z') \frac{đ z'}{z' - z''} + \oint X(z') \frac{đ z'}{z' - 1} \right] . \tag{2.6}$$

Substituting (2.3) in (2.6) yields

$$\pi_- X(z) = X_-(z) + (X_+(1) + X_0 - X_-(1))/2 \tag{2.7}$$

where

$$X_+(z) = \sum_{j>0} X_j z^j \tag{2.8}$$

and

$$X_-(z) = \sum_{j<0} X_j z^j . \tag{2.9}$$

Similarly we find

$$\pi_+ X(z) = X_+(z) + (X_-(1) + X_0 - X_+(1))/2 . \tag{2.10}$$

Formulae (1.35) and (1.36), and the fact that $\pi_\pm$ map $\underline{g} \to \underline{g}$, are now immediate from these formulae. From (1.36), for $X, Y \in \underline{g}$, $\pi_-[\pi_+ X, \pi_+ Y]$ is a constant given by $([\pi_+ X, \pi_+ Y](1))/2 = [X(1), Y(1)]/4 = 0$, as both matrices are diagonal. Similarly $\pi_+[\pi_- X, \pi_- Y] = 0$. Now $R = \pi_+ - \pi_-$, so

$$\pi_+ = (1 + R)/2 , \qquad \pi_- = (1 - R)/2 , \tag{2.11}$$

which lead to the formulae

$$\begin{aligned} (1 - R)[(1 + R)X,\, (1 + R)Y] &= 0 , \\ (1 + R)[(1 - R)X,\, (1 - R)Y] &= 0 . \end{aligned} \tag{2.12}$$

Adding and multiplying out yields

$$[RX, RY] - R([RX, Y] + [X, RY]) = -[X, Y] \tag{2.13}$$

which is the modified Yang-Baxter equation. Thus $[X,Y]_R$ following (1.37) is a Lie-bracket,and the set $\underline{g}$ together with the structure $[\cdot,\cdot]_R$ forms a Lie-algebra which we denote by $\tilde{\underline{g}}$ as in Section 1 (note that $[X,Y]_R(z=1)=0$, which is diagonal).

Define

$$\begin{aligned}\widetilde{G} = \{g \in G : g &\text{ has a factorization } g(z) = g_+(z)g_-(z), \text{ where the invertible} \\ &\text{loops } g_\pm(\cdot) \text{ have invertible analytic continuations to} \\ &\{|z|<1\}\ ,\ \{1<|z|\leq\infty\}\text{respectively, which are smooth up to} \\ &\text{the boundary and contractible (preserving analyticity) to the} \\ &\text{identity. Also } g_+(1) = g_-(1) = \sqrt{g(1)} > 0\}\ .\end{aligned} \tag{2.14}$$

Note that the factorization $g_\pm$, if it exists, is unique. Indeed if $g_+g_- = g'_+g'_-$ are two such factorizations, then $(g'_+(z))^{-1}g_+(z) = g'_-(z)(g_-(z))^{-1}$ is bounded and analytic in the entire $z$ plane, and hence is constant. Evaluating at $z=1$ we see that the constant must be the identity. As $g(z)=\overline{g(\bar{z})}$, uniqueness implies

$$g_\pm(z) = \overline{g_\pm(\bar{z})} \tag{2.15}$$

Note also that the requirement of contractibility, preserving analyticity, is not independent and follows by a standard argument from the other properties of $g(z)$ in the definition of $\widetilde{G}$.

For $g,h\in\widetilde{G}$, define the multiplication

$$g * h \equiv g_+hg_- = (g_+h_+)(h_-g_-) \tag{2.16}$$

One checks directly that $(\widetilde{G},*)$ is a group. Note that the $*$-inverse of $g=g_+g_-$ is $g_+^{-1}g_-^{-1}$. For $X\in\underline{g}$, $e^{t\pi_+X}\,e^{t\pi_-X}$ is a curve in $\widetilde{G}$ with derivative $\pi_+X+\pi_-X=X$ at the identity. Thus

$$\begin{aligned}\widetilde{\mathrm{Ad}}_gX &= \frac{d}{dt}\Big|_{t=0} g*e^{t\pi_+X}*g^{-1} \\ &= \frac{d}{dt}\Big|_{t=0} g_+e^{t\pi_+X}g_+^{-1}g_-^{-1}e^{t\pi_-X}g_- \\ &= g_+(\pi_+X)g_+^{-1}+g_-^{-1}(\pi_-X)g_-\ ,\end{aligned} \tag{2.17}$$

and hence

$$\begin{aligned}\widetilde{[Y,X]} &= \frac{d}{dt}\Big|_{t=0} \widetilde{\mathrm{Ad}}_{e^{t\pi_+ Y} e^{t\pi_- Y}} X \\ &= [\pi_+ Y, \pi_+ X] - [\pi_- Y, \pi_- X] \\ &= \frac{1}{2}([RY, X] + [Y, RX]) \\ &= [Y, X]_R \,. \end{aligned} \tag{2.18}$$

Thus $\underline{\tilde{g}}$ is the Lie-algebra of $\widetilde{G}$.

**(b) Dual Lie-algebras and coadjoint orbits**

Notice first that under (2.1) the loop $\widetilde{M}(\lambda) = \dfrac{1 - \lambda M - \lambda^2 J^2}{1 - \lambda^2}$ becomes $\frac{A_{-1}}{z} + A_0 + A_1 z$, where $A_1 = \frac{J^2 - M - 1}{4}$, $A_0 = \frac{1 + J^2}{2}$ and $A_{-1} = A_1^T = \frac{J^2 + M - 1}{4}$. One would hope that elements of the form $A_{-1}z^{-1} + A_0 + A_1 z$ generated a coadjoint orbit consisting of elements of the same form. As we will see, however, this is not the case and generically elements of the form $A_{-1}z^{-1} + A_0 + A_1 z$ generate orbits consisting of elements of the form $\dfrac{A_p}{z-1} + A_{-1}z^{-1} + A_0 + A_1 z$ with a polar singularity at $z = 1 \in S^1$.

Motivated by these considerations, we define

$$\begin{aligned}\underline{\tilde{g}}^*_{\text{sing}} \equiv \{A : A(z) = \frac{A_p}{z-1} + A_{\text{reg}}(z) = \overline{A(\bar{z})} \in g\ell(N, \mathbb{C}) \text{ and} \\ A_{\text{reg}}(\cdot) \text{ is a smooth loop on } S^1\} \,.\end{aligned} \tag{2.19}$$

In an obvious notation

$$\underline{\tilde{g}}^*_{\text{sing}} = \underline{\tilde{g}}^*_{\text{pole}} + \underline{\tilde{g}}^*_{\text{reg}} \,. \tag{2.20}$$

We will show below that $\underline{\tilde{g}}^*_{\text{sing}}$ is invariant under the coadjoint action of $\widetilde{G}$, and the coadjoint orbits of interest will be found to be subsets of $\underline{\tilde{g}}^*_{sing}$.

Elements $A$ in $\underline{\tilde{g}}^*_{\text{sing}}$ induce linear functionals on $\underline{\tilde{g}}$ through the non-degenerate pairing

$$(A, X) \equiv ⨍ \operatorname{tr} A(z) X(z) \frac{\bar{d} z}{z} \tag{2.21}$$

Note that on $\underline{g} \times \underline{g}$ this pairing is ad-invariant,

$$(X, [Y, Z]) = -([Y, X], Z) \,. \tag{2.22}$$

Suppose $A = \sum_j A_j z^j \in \underline{\tilde{g}}^*_{\text{reg}}$. We compute $\pi_+^* A$. For $X \in \underline{\tilde{g}}$,

$$
\begin{aligned}
(\pi_+^* A, X) &\equiv (A, \pi_+ X) \\
&= \int \operatorname{tr} A(z)(\pi_+ X)(z)\frac{dz}{z} \\
&= \operatorname{tr}(\pi_+ A)(0)(\pi_+ X)(0) + (\pi_- A, X) - \operatorname{tr}(\pi_- A)(\infty)(\pi_- X)(\infty) \, , \quad (2.23)
\end{aligned}
$$

where we have used Cauchy's theorem repeatedly. But for any constant matrix $C$,

$$
\oint \operatorname{tr} CX(z)\frac{đz}{(z-1)z} = -\operatorname{tr} C(\pi_+ X)(0) \tag{2.24}
$$

and

$$
\oint \operatorname{tr} CX(z)\frac{đz}{z-1} = \operatorname{tr} C(\pi_- X)(\infty) \, . \tag{2.25}
$$

Applying these formulae to (2.23), we find

$$
\begin{aligned}
\pi_+^* A &= \pi_- A - \frac{(\pi_+ A)(0)}{z-1} + \frac{z(\pi_- A)(\infty)}{z-1} \\
&= A_- - \frac{A_0}{z-1} \, .
\end{aligned} \tag{2.26}
$$

Now suppose $A = \dfrac{A_p}{z-1} \in \underline{\tilde{g}}^*_{\text{pol}}$. Then a similar calculation shows that

$$
\pi_+^* A = \frac{A_p}{z-1} + \frac{1}{4}(A_p)_{\text{diag}}\delta_1 \, , \tag{2.27}
$$

where $(A_p)_{\text{diag}}$ is the diagonal part of $A_p$ and $\delta_1$ is the delta function at $z = 1$,

$$
\int_{|z|=1} B(z)\delta_1(z)\, đz = B(1) \, .
$$

Note that as $X(1)$ is diagonal, only the diagonal part of $A_p$ plays a role at $z = 1$ in (2.27).

We have proved the following result:

**Lemma 2.28.** *For $A = A_p/(z-1) + A_{\text{reg}} \in \underline{\tilde{g}}^*_{\text{sing}}$,*

$$
\pi_+^* A = \frac{A_p - (A_{\text{reg}})_0}{z-1} + (A_{\text{reg}})_- + \frac{1}{4}(A_p)_{\text{diag}}\delta_1 \tag{2.29}
$$

and

$$\pi_-^* A = \frac{(A_{\rm reg})_0}{z-1} + (A_{\rm reg})_+ + (A_{\rm reg})_0 - \frac{1}{4}(A_p)_{\rm diag}\delta_1 \ . \tag{2.30}$$

□

**Remark.** The use of the symbols $\pi_\pm A$ above involves a slight but irrelevant abuse of notation as $A(1)$ need not be diagonal, and so $A$ may not lie in $\underline{g}$ where $\pi_\pm$ are defined.

For convenience in this subsection we replace $g_-$ by $g_-^{-1}$ in the basic factorization for $g \in \widetilde{G}$. Thus

$$g = g_+ g_-^{-1} \tag{2.31}$$

and (2.17) takes the more symmetrical form

$$\widetilde{\rm Ad}_g X = g_+(\pi_+ X)g_+^{-1} + g_-(\pi_- X)g_-^{-1} \ . \tag{2.32}$$

In *all* the other parts of the paper we retain the factorization as in (2.14), which is more convenient for the analytical questions that arise.

For $A \in \tilde{\underline{g}}^*_{\rm sing}$, and $g \in \widetilde{G}$, $g_\pm^{-1} A g_\pm$ also lie in $\tilde{\underline{g}}^*_{\rm sing}$, and from (2.32)

$$\widetilde{\rm Ad}_g^* A = \pi_+^*(g_+^{-1} A g_+) + \pi_-^*(g_-^{-1} A g_-) \tag{2.33}$$

**Lemma 2.34.** *$\widetilde{\rm Ad}_g^*$ maps $\tilde{\underline{g}}^*_{\rm sing}$ into itself. Moreover for any $k, \ell \in \mathbb{Z}_+$,* the set

$$\{A : A = \frac{A_p}{z-1} + \frac{A_{-k}}{z^k} + \cdots + A_\ell z^\ell\} \text{ is } \widetilde{\rm Ad}_g^*\text{-invariant.} \tag{2.35}$$

**Proof:** To prove $\widetilde{\rm Ad}_g^* \tilde{\underline{g}}^*_{\rm sing} \subset \tilde{\underline{g}}^*_{\rm sing}$ it is enough by Lemma 2.28 to show that

$$\big((g_+^{-1} A g_+)_p\big)_{\rm diag} - \big((g_-^{-1} A g_-)_p\big)_{\rm diag} = 0 \tag{2.36}$$

But

$$\big((g_\pm^{-1} A g_\pm)_p\big)_{\rm diag} = \big(g_\pm^{-1}(1) A_p \, g_\pm(1)\big)_{\rm diag} = (A_p)_{\rm diag} \ ,$$

as $g_\pm(1)$ are diagonal, which proves (2.36).

Now take $A = \frac{A_p}{z-1} + \frac{A_{-k}}{z^k} + \cdots + A_\ell z^\ell$. From the analyticity properties of $g_\pm$, it follows that

$$g_+^{-1} A g_+ = \frac{B_p}{z-1} + \sum_{j=-k}^{\infty} B_j z^j \tag{2.37}$$

and

$$(g_-^{-1} A g_-) = \frac{C_p}{z-1} + \sum_{-\infty}^{\ell} C_j z^j \tag{2.38}$$

for suitable matrix coefficients $B_p, B_j, C_p, C_j$. Hence

$$\left((g_+^{-1} A g_+)_{\text{reg}}\right)_- = \sum_{j=-k}^{-1} B_j z^j$$

and

$$\left((g_-^{-1} A g_-)_{\text{reg}}\right)_+ = \sum_{j=1}^{\ell} C_j z^j \ .$$

The result now follows from (2.29) and (2.30). $\square$

In the remainder of this subsection we consider only the case $k = \ell = 1$,

$$A = \frac{A_p}{z-1} + \frac{A_{-1}}{z} + A_0 + A_1 z \ .$$

The above formulae reduce to

$$(\widetilde{\mathrm{Ad}}_g^* A)(z) = \frac{B_p - B_0 + C_0}{z-1} + B_{-1} z^{-1} + C_0 + C_1 z \ ,$$

and a straightforward computation of the constants yields, finally, the formula

$$\begin{aligned}
(\widetilde{\mathrm{Ad}}_g^* A)(z) = {} & \frac{1}{z-1}\Big\{ \left[g_+(0)^{-1} g_+'(0), g_+(0)^{-1} A_{-1} g_+(0)\right] + g_+(0)^{-1}(A_p - A_0) g_+(0) \\
& - \left[g_-(\infty)^{-1} g_-'(\infty), g_-(\infty)^{-1} A_1 g_-(\infty)\right] + g_-(\infty)^{-1} A_0 g_-(0)\Big\} \\
& + \frac{g_+(0)^{-1} A_{-1} g_+(0)}{z} \\
& - \left[g_-(\infty)^{-1} g_-'(\infty), g_-(\infty)^{-1} A_1 g_-(\infty)\right] + g_-(\infty)^{-1} A_0 g_-(\infty) \\
& + g_-(\infty)^{-1} A_1 g_-(\infty) z \ ,
\end{aligned} \tag{2.39}$$

where

$$g_+'(0) = \frac{dg_+}{dz}(z=0) \quad \text{and} \quad g_-'(\infty) = \frac{dg_-}{dz^{-1}}(z=\infty) \ . \tag{2.40}$$

Our goal is to show (see Corollary to Theorem 2.76 below) that for generic $A_p$, $A_{-1}$, $A_0$ and $A_1$, the coadjoint orbit

$$O_A = \{\widetilde{\mathrm{Ad}}_g^*(\frac{A_p}{z-1} + \frac{A_{-1}}{z} + A_0 + A_1 z) : g \in \widetilde{G}\} \tag{2.41}$$

has dimension $4N^2 - 4N$. But first we consider briefly an abstract question. Let $G_1$ and $G_2$ be Lie-groups, with Lie-algebras $\underline{g}_1$ and $\underline{g}_2$ and dual Lie-algebras $\underline{g}_1^*$ and $\underline{g}_2^*$, respectively. Let $\Phi$ be a homomorphism from $G_1$ to $G_2$, with derivative $\phi$ at the identity,

$$\begin{aligned} \phi &= \Phi'(e_1) : \underline{g}_1 \to \underline{g}_2 \,, \\ \phi(x) &= \frac{d}{dt}\bigg|_{t=0} \Phi(e^{tx}) \,. \end{aligned} \tag{2.42}$$

We say that $F : \underline{g}^* \to \mathbb{C}$ is *smooth on* $\underline{g}^*$ if $F$ is smooth with derivative $dF(\alpha) \in \underline{g}$ for all $\alpha \in \underline{g}^*$. In finite dimensions the derivative trivially belongs to $\underline{g}$; in infinite dimensions this is an additional assumption. Recall that $dF(\alpha)$ is defined as a linear functional on $\underline{g}^*$ through

$$\frac{d}{dt}\bigg|_{t=0} F(\alpha + \beta t) = dF(\alpha)(\beta)$$

for all $\beta \in \underline{q}^*$.

**Lemma 2.43.**

(i) *$\phi$ is a Lie-algebra homomorphism from $\underline{g}_1$ to $\underline{g}_2$.*

(ii) *$\phi^*$ is a Poisson map from $\underline{g}_2^*$ to $\underline{g}_1^*$ equipped with the Lie-Poisson structures.*

(iii) $\mathrm{Ad}_{\Phi(g)} \circ \phi = \phi \circ \mathrm{Ad}_g$.

(iv) $\phi^* \circ \mathrm{Ad}_{\Phi(g)}^* = \mathrm{Ad}_g^* \circ \phi^*$.

**Proof:** This lemma is standard. We prove only (ii). For $F$, $G$ smooth on $\underline{g}_1^*$, we must show that

$$\{F \circ \phi^*, G \circ \phi^*\} = \{F, G\} \circ \phi^* \,, \tag{2.44}$$

or

$$\alpha\big([d(F \circ \phi^*)(\alpha) \,, d(G \circ \phi^*)(\alpha)]\big) = \phi^*(\alpha)\big([dF(\phi^*(x)) \,, dG(\phi^*(\alpha))]\big)$$

for all $\alpha \in \underline{g}_2^*$. But for $\alpha, \beta \in \underline{g}_2^*$,

$$\begin{aligned} d(F \circ \phi^*)(\alpha)(\beta) &= \frac{d}{dt}\Big|_{t=0} F \circ \phi^*(\alpha + t\beta) \\ &= dF(\phi^*(\alpha))(\phi^*(\beta)) \\ &= \phi(dF(\phi^*(\alpha)))(\beta) . \end{aligned}$$

Thus

$$d(F \circ \phi^*)(\alpha) = \phi(dF(\phi^*(\alpha))) \in \underline{g}_1 ,$$

and hence $F \circ \phi^*$ is smooth on $\underline{g}_2^*$. We now have

$$\begin{aligned} \alpha([d(F \circ \phi^*)(\alpha) , d(G \circ \phi^*)(\alpha)]) &= \alpha([\phi(dF(\phi^*(\alpha))), \phi(dG(\phi^*(\alpha))]) \\ &= \alpha(\phi([dF(\phi^*(\alpha)), dG(\phi^*(\alpha))])) \\ &= \phi^*(\alpha)([dF(\phi^*(\alpha)) , dG(\phi^*(\alpha))]) , \end{aligned}$$

as desired. □

The point is this. From (iv) we have $O_{\phi^*(\alpha)} \subset \phi^*(O_\alpha)$ for all $\alpha \in \underline{g}_2^*$, and if $\Phi$ is onto $G_2$, then

$$O_{\phi^*(\alpha)} = \phi^*(O_\alpha) . \tag{2.45}$$

Moreover, if $\phi$ is onto $\underline{g}_2$, then $\phi^*$ is injective and hence bijective from $O_{\phi^*(\alpha)}$ onto $O_\alpha$; in particular they have the same dimension. We will construct a homeomorphism $\Phi$ from $G_1 = \widetilde{G}$ onto a finite dimensional Lie-group $G_2$, with $\phi$ surjective, for which the finite dimensional orbits $O_{\phi^*(\alpha)}$ given by (2.45) are precisely the orbits $O_A$ in (2.41). In this way the computation of the orbits is reduced to an equivalent problem for a finite dimensional group.

Let $G_0$ denote the identity component of the direct sum of two copies of the semi-direct product $G\ell(N,\boldsymbol{R})_{ad} \times g\ell(N,\boldsymbol{R})$,

$$G_0 = \{((g,u),(h,v)) : g,h \in G\ell(N,\boldsymbol{R}),\ \det g > 0,\ \det h > 0 ,\ u,v \in g\ell(N,\boldsymbol{R})\} , \tag{2.46}$$

$$((g,u),(h,v)) \circ ((g',u'),(h',v')) = ((g'g, u' + g'u(g')^{-1}) , (h'h, v' + h'v(h')^{-1})) . \tag{2.47}$$

Note that $((I_N, 0), (I_N, 0))$ is the identity in $G_0$ and

$$((g, u), (h, v))^{-1} = (((g^{-1}, -g^{-1}ug), (h^{-1}, -h^{-1}vh)). \tag{2.48}$$

The Lie-algebra $\underline{g}_0$ of $G_0$ is given by

$$\underline{g}_0 = \{((K, L), (U, V)) : K, L, U, V \in g\ell(N, \mathbf{R})\} \tag{2.49}$$

with Lie-bracket

$$\begin{aligned} &[((K, L), (U, V)), ((K', L'), (U', V'))] \\ &= (([K', K], [K', L] + [L', K]), ([U', U], [U', V] + [V', U])). \end{aligned} \tag{2.50}$$

Define the map $\Phi : \widetilde{G} \to G_0$ by

$$\Phi(g) = (((g_+(0))^{-1}, (g_+(0))^{-1}g'_+(0)), (g_-(\infty)^{-1}, g_-(\infty)^{-1}g'_-(\infty))), \tag{2.51}$$

where $g = g_+g_-^{-1}$ and $g'_+(0)$, $g'_-(\infty)$ are defined in (2.40). Note that $g_+(t)$ is real for $0 \le t \le 1$ by (2.15). As $g_+(z)$ is invertible for all $|z| \le 1$, and as $g_+(1) > 0$ (see (2.14)), it follows that $\det g_+(0)$, and similarly $\det g_-(\infty)$, is positive.

The map $\Phi$ is a surjective homomorphism. Indeed for $g_1 = g_{1+}g_{1-}^{-1}$, $g_2 = g_{2+}g_{2-}^{-1}$,

$$\begin{aligned} \Phi(g_1) \circ \Phi(g_2) &= ((g_{1+}(0)^{-1}, g_{1+}(0)^{-1}g'_{1+}(0)), (g_{1-}(\infty)^{-1}, g_{1-}(\infty)^{-1}g'_{1-}(\infty))) \\ &\quad \circ ((g_{2+}(0)^{-1}, g_{2+}(0)^{-1}g'_{2+}(0)), (g_{2-}(\infty)^{-1}, g_{2-}(\infty)^{-1}g'_{2-}(\infty))) \\ &= (g_{2+}(0)^{-1}g_{1+}(0)^{-1}, g_{2+}(0)^{-1}g'_{2+}(0) \\ &\quad + g_{2+}(0)^{-1}g_{1+}(0)^{-1}g'_{1+}(0)g_{2+}(0)), \\ &\quad (g_{2-}(\infty)^{-1}g_{1-}(\infty)^{-1}, g_{2-}(\infty)^{-1}g'_{2-}(\infty) \\ &\quad + g_{2-}(\infty)^{-1}g_{1-}(\infty)^{-1}g'_{1-}(\infty)g_{2-}(\infty)). \end{aligned} \tag{2.52}$$

But $g_1 * g_2 = g_{1+}g_2g_{1-}^{-1} = g_{1+}g_{2+}(g_{1-}g_{2-})^{-1}$ and so

$$\begin{aligned} \Phi(g_1 * g_2) = (&((g_{1+}g_{2+})^{-1}(0), (g_{1+}g_{2+})^{-1}(0)(g'_{1+}(0)g_{2+}(0) + g_{1+}(0)g'_{2+}(0))), \\ &((g_{1-}g_{2-})^{-1}(\infty), (g_{1-}g_{2-})^{-1}(\infty)(g'_{1-}(\infty)g_{2-}(\infty) + g_{1-}(\infty)g'_{2-}(\infty)))), \end{aligned}$$

which agrees with (2.52), verifying the homomorphism property. To show that $\Phi$ is onto, suppose $((M,K),(X,Y)) \in G_0$, det $M$, det $X > 0$. We must show that there exists $g = g_+ g_-^{-1} \in \widetilde{G}$ such that $\Phi(g) = ((M,K),(X,Y))$. Recall that we must have $g(z) = \overline{g(\bar{z})}$ and $g(\cdot)$ homotopic to the identity. By the homomorphism property, and by the symmetry between $z = 0$ and $z = \infty$, it is enough to show that there exists $g = g_+(z)$ (necessarily $g_+(1) = g_-(1) = I$) such that

$$(g_+(0)^{-1}, g_+(0)^{-1} g_+'(0)) = (M,K) = (M,0) \circ (I,K) , \tag{2.53}$$

and again by the homomorphism property it is enough to consider $(M,0)$ and $(I,K)$ separately.

For $(I,K)$, a suitable loop is

$$g = g_+ = e^{z(1-z)K} , \tag{2.54}$$

with homotopy

$$g_t = (g_t)_+ = e^{tz(1-z)K} , \quad 0 \le t \le 1 . \tag{2.55}$$

For $(M,0)$ with det $M > 0$, it is sufficient by the homomorphism property and connectivity to consider the case

$$M = (I + \epsilon)^{-1} \tag{2.56}$$

where the real matrix $\epsilon$ is small. But here

$$g = g_+ = I + \epsilon(1 - z^2) \tag{2.57}$$

is an appropriate loop, deformable to the identity via

$$g_t = (g_t)_+ = I + t\epsilon(1 - z^2) , \qquad 0 \le t \le 1 . \tag{2.58}$$

This completes the proof that $\Phi$ is surjective.

For $X = \sum_{j=-\infty}^{\infty} X_j z^j = \pi_+ X + \pi_- X \in \tilde{\underline{g}}$, a simple computation shows that

$$\begin{aligned}\phi(X) &= \frac{d}{dt}\bigg|_{t=0} \Phi(e^{t\pi_+ X} e^{t\pi_- X}) \\ &= ((-(\pi_+ X)(0), X_1), ((\pi_- X)(\infty), -X_{-1})) . \end{aligned} \tag{2.59}$$

Also it is easy to see that $\phi$ is onto $\underline{g}_0$ (for this purpose it is enough to consider $X$ of the form $\frac{X_{-2}}{z^2} + \cdots + X_2 z^2$). Thus $\phi^*$ is injective from $\underline{g}_0^*$ into $\tilde{\underline{g}}^*$.

To compute $\phi^*$ we identify $\underline{g}_0^*$ with $\underline{g}_0$ through the standard nondegenerate pairing,

$$(((M,K),(R,S)), ((M',K'),(R',S'))) = \text{ tr } MM' + \text{ tr } KK' + \text{ tr } RR' + \text{ tr } SS' . \tag{2.60}$$

Then for $\alpha = ((M,K), (R,S)) \in \underline{g}_0^* \cong \underline{g}_0$,

$$\begin{aligned}\phi^*\alpha(X) &= (\alpha, \phi(x)) \\ &= \text{ tr } KX_1 - \text{ tr } M\pi_+ X(0) - \text{ tr } SX_{-1} + \text{ tr } R\pi_- X(\infty) .\end{aligned}$$

On the other hand, for $\beta = B_p/(z-1) + B_{-1}z^{-1} + B_0 + B_1 z$, a now familiar computation shows that

$$(\beta, X) = \text{ tr}(B_0 - B_p)\pi_+ X(0) + \text{ tr } B_{-1}X_1 + \text{ tr } B_0\pi_-(\infty) + \text{ tr } B_1 X_{-1} .$$

We conclude that

$$\phi^*((M,K), (R,S)) = \frac{R+M}{z-1} + \frac{K}{z} + R - zS , \tag{2.61}$$

or alternatively,

$$\phi^*(((B_p - B_0), B_{-1}), (B_0, -B_1)) = \frac{B_p}{z-1} + \frac{B_{-1}}{z} + B_0 + B_1 z . \tag{2.62}$$

Thus the orbits through points $\phi^*\alpha \in \tilde{\underline{g}}^*$ are precisely the orbits considered in (2.41). To compute the dimension of the coadjoint orbit through $\phi^*\alpha$, it is sufficient, by Lemma 2.43, to compute the dimension of the orbit through $\alpha$ in $\underline{g}_0^*$. A standard computation shows that for $((g,u),(h,v)) \in G_0$

$$\begin{aligned}&\mathrm{Ad}^*_{((g,u),(h,v))}((M,K),(R,S)) \\ &= ((gMg^{-1} + [u, gKg^{-1}], gKg^{-1}), (hRh^{-1} + [v, hSh^{-1}], hSh^{-1})) ,\end{aligned} \tag{2.63}$$

and we can consider each pair $(M, K)$ and $(R, S)$ separately.

First observe that

$$\phi_{1k}(M,K) = \phi_{1k}(K) = \text{ tr } K^k \ , \quad \phi_{2k}(M,K) = \text{ tr } MK^k \ , \quad k = 1,\ldots,N \ , \tag{2.64}$$

are coadjoint invariants. Indeed

$$\begin{aligned} &\operatorname{tr}(gMg^{-1} + [u, gKg^{-1}])(gKg^{-1})^k \\ &= \text{ tr } MK^k + \operatorname{tr}(u[gKg^{-1}, gK^kg^{-1}]) = \text{ tr } MK^k \ . \end{aligned}$$

**Lemma 2.65.** *The functions $\phi_{1k}, \phi_{2k}$, $1 \le k \le N$, are independent at $(M,K)$ if $K$ is invertible with distinct eigenvalues. If* $\det K = 0$*, then the functions are dependent at* $(M,K)$.

**Proof:** Suppose that $\sum_{k=1}^N a_k \frac{\partial \phi_{1k}}{\partial K} + \sum_{k=1}^N b_k \frac{\partial \phi_{2k}}{\partial K} = 0$, $\sum_{k=1}^N a_k \frac{\partial \phi_{1k}}{\partial M} + \sum_{k=1}^N b_k \frac{\partial \phi_{2k}}{\partial M} = 0$ for suitable constants $a_k, b_k$. Then as $\partial\phi_{1k}/\partial M = 0$, we obtain $\sum_{k=1}^N b_k (K^T)^k = 0$, or $\sum_{k=1}^N b_k K^{k-1} = 0$, as $K$ is invertible. But this is impossible as $K$ has distinct eigenvalues, unless $b_k = 0$, $k = 1,\ldots,n$. The first equation now implies $\sum_{k=1}^N a_k k K^{k-1} = 0$, and again $a_k = 0$, $k = 1,\ldots,n$. On the other hand, if $\det K = 0$, then $P_K(K) = 0$, where $P_K(\lambda) = \sum_{j=0}^N \sigma_j \lambda^j$ is the characteristic polynomial of $K$. But $\sigma_0 = \det K = 0$, and so $(\{a_j = 0\}_{j=1}^N, \{b_j = \sigma_j\}_{j=1}^N) \neq 0$ solves the above gradient equations for $\phi_{1k}$ and $\phi_{2k}$. This completes the proof of the lemma. □

**Definition 2.66.** We say that a pair $(M,K)$ is *generic* if $K$ is invertible with distinct eigenvalues.

Clearly generic pairs $(M,K)$ form a dense, open set in $\boldsymbol{R}^{2N^2}$. Given a generic pair $(M_0, K_0)$, consider

$$S_{(M_0,K_0)} = \{(M,K) : \phi_{ik}(M,K) = \phi_{ik}(M_0,K_0) \ , \ 1 \le i \le 2, 1 \le k \le N\} \ . \tag{2.67}$$

Then $S_{(M_0,K_0)}$ is a submanifold of dimension $2N^2 - 2N$ in $\boldsymbol{R}^{2N^2}$. Indeed if $(M,K) \in S_{(M_0,K_0)}$, then spec $K =$ spec $K_0$ and so $(M,K)$ is generic and hence $d\phi_{1k}$ are independent at $(M,K)$. Clearly

$$O_{(M_0,K_0)} \subset S_{(M_0,K_0)} \ . \tag{2.68}$$

We will see that $O_{(M_0,K_0)} = S_{(M_0,K_0)}$ in all cases except the case where $N$ is even and $K$ has *no* real eigenvalues; in this case $O_{(M_0,K_0)}$ is an appropriate component of $S_{(M_0,K_0)}$, as we show below (see Theorem 2.76).

So suppose $(M_0, K_0)$ is generic. Define

$$S_{K_0} \equiv \{K : K \text{ is a real matrix with spec } K = \text{ spec } K_0\} . \tag{2.69}$$

By elementary linear algebra

$$S_{K_0} = S_{K_0,+} \cup S_{K_0,-} \tag{2.70}$$

where

$$\begin{aligned} S_{K_0,\pm} = \{gK_0g^{-1} : g \text{ is a real matrix with } \det g > 0 ,\\ \det g < 0, \text{ respectively}\} . \end{aligned} \tag{2.71}$$

There are two cases:

$$\left.\begin{array}{ll} (1) & \text{spec } K_0 \cap R \neq \emptyset \\ (2) & \text{spec } K_0 \cap R = \emptyset \end{array}\right\} . \tag{2.72}$$

(Of course, case (2) can only occur if $N$ is even.) In case (1), a simple compuation using the real Jordan form (recall spec $K_0$ is simple), shows that

$$\left.\begin{aligned} S_{K_0,+} &= S_{K_0,-} \\ &= S_{K_0} \end{aligned}\right\} . \tag{2.73}$$

In case (2)

$$S_{K_0,+} \cap S_{K_0,-} = \emptyset . \tag{2.74}$$

This in turn follows from the observation that a matrix commuting with a real Jordan form $J = \text{diag}(B_1, \ldots, B_{N/2})$, where the $B_i$ are all $2 \times 2$ blocks with distinct spectrum, must also have the same block structure and positive determinant.

In case (2), this implies that there exists a continuous map $P : S_{K_0} \to \{\pm 1\}$ with the property that for $K, K' \in S_{K_0}$,

$$K = gK'g^{-1} , \quad \det g > 0 \Leftrightarrow P(K) = P(K') . \tag{2.75}$$

We leave it to the reader to verify that if $K = gJg^{-1}$ with deg $g > 0$ and $J$ in real Jordan form, then $P(K)$ can be taken to be the signature of the product of the nonzero, strictly upper-triangular elements of $J$. In particular if $K$ is skew, then $P(K)$ is the sign of the Pfaffian of $K$.

If $(M, K) \in O_{(M_0, K_0)}$, then clearly $K \in S_{K_0,+}$. Conversely suppose $(M, K) \in S_{(M_0,K_0)}$ with $K \in S_{K_0,+}$, so that $K = gK_0 g^{-1}$ with deg $g > 0$. Since tr $MK^\ell =$ tr $M_0 K_0^\ell$, $\ell = 1, \ldots, N$, we have that $g^{-1}Mg - M_0$ is orthogonal, under the pairing $(A, B) =$ tr $AB$, to the algebra generated by $K_0$, which is equivalent to saying that $g^{-1}Mg - M_0 = [\omega, K_0]$, for some real $\omega$, as $K_0$ has simple spectrum. For $u = g\omega g^{-1}$, we then have $M = gM_0 g^{-1} + [u, gK_0 g^{-1}]$. Thus $(M, K) = Ad^*_{(g,u)}(M_0, K_0)$ with deg $g > 0$.

The above calculations, together with their analogs for $(R, S)$, yield the following result. We say that $((M_0, K_0), (R_0, S_0))$ is *generic* if both pairs are generic.

**Theorem 2.76.** *Suppose* $((M_0, K_0), (R_0, S_0))$ *is generic. Then* $O_{((M_0,K_0),(R_0,S_0))}$ *is a* $4N^2 - 4N$ *dimensional manifold given by*

$$
\begin{aligned}
O_{((M_0,K_0),(R_0,S_0))} = \{((M, K), (R, S)) : \; & \text{tr } MK^k = \text{ tr } M_0 K_0^k, \text{ tr } K^k = \text{ tr } K_0^k, \\
& \text{tr } RS^k = \text{ tr } R_0 S_0^k, \text{ tr } S^k = \text{ tr } S_0^k, \; 1 \le k \le N, \\
& \text{and in the case that } N \text{ is even, and} K_0 \text{ and/or} \\
& S_0 \text{ has no real eigenvalues, then, in addition,} \\
& P(K) = P(K_0) \text{ and/or } P(S) = P(S_0)\} \; . 
\end{aligned} \tag{2.77}
$$

From (2.62) we see that loops $\dfrac{A_p}{z-1} + \dfrac{A_{-1}}{z} + A_0 + A_1 z$ for which $A_1$ and $A_{-1}$ are invertible with distinct eigenvalues, generate coadjoint orbits which are $4N^2 - 4N$ dimensional. We will also call such loops *generic* (within the class of such loops).

**Corollary to Theorem 2.76.** *Suppose* $A^0 = \dfrac{A_p^0}{z-1} + \dfrac{A_{-1}^0}{z} + A_0^0 + A_1^0 z$ *is generic.*

*Then $O_{A^0}$ is a $4N^2 - 4N$ dimensional manifold given by*

$$O_{A^0} = \{A = \frac{A_p}{z-1} + \frac{A_{-1}}{z} + A_0 + A_1 z : \ \mathrm{tr}\ A_1^k = \mathrm{tr}(A_1^0)^k,\ \mathrm{tr}\ A_0 A_1^k = \mathrm{tr}\ A_0^0 (A_1^0)^k,$$

$$\mathrm{tr}\ A_{-1}^k = \mathrm{tr}(A_{-1}^0)^k,\ \mathrm{tr}(A_p - A_0)A_{-1}^k$$

$$= \mathrm{tr}(A_p^0 - A_0^0)(A_{-1}^0)^k,\ 1 \le k \le N, \text{ and in the}$$

case that $N$ is even, and $A_1^0$ and/or $A_{-1}^0$

has no real eigenvdalues, then, in addition,

$$P(A_1) = P(A_1^0) \text{ and/or}$$

$$P(A_{-1}) = P(A_{-1}^0)\}. \tag{2.78}$$

□

**Remark.** The reduction from $\tilde{g}^*$ to the finite dimensional dual Lie-algebra $\underline{g}_0^*$, does not simplify the *dynamical* problem. In fact it is harder to solve the flows of interest on $\underline{g}_0^*$ than on $\tilde{g}^*$. This is because the $R$-matrix structure is lost and the solution by factorization is no longer at hand. As in QR (see [DLT], [DL]) the precise opposite is in fact the point: one should lift the flows from the finite dimensional dual Lie-algebra to a loop algebra in order to solve the problem.

**(c) Commuting integrals on a generic orbit** $O_A = O_{A_p/(z-1)+A_{-1}z^{-1}+A_0+A_1 z}$.

In this section we construct $\frac{1}{2}(4N^2 - 4N) = 2N^2 - 2N$ commuting functionals on a generic orbit $O_A$. As we will see (subsection (d) below), these functionals provide integrals for the flow that interpolates the Moser-Veselov algorithm.

From (1.39) we see that for the discrete Euler-Arnold equation the curve

$$\det(M(t,\lambda) - \eta) = 0$$

is preserved in time. For $A \in \underline{g}_{sing}^*$, this leads us to consider the curve $\det((z-1)A(z) - \eta) = 0$ with coefficients

$$I_{rk}(A) = \int_{|z|=1} \int_{|\eta|=1} \det((z-1)A(z) - \eta) \frac{d\!\!\!{}^-\eta}{\eta^{n-r+1}} \frac{d\!\!\!{}^-z}{z^{k+1}} \tag{2.79}$$

for all $r, k \in \mathbb{Z}$.

**Theorem 2.80.** *The $I_{rk}$'s are a family of smooth, Poisson commuting functions on $\underline{\tilde{g}}^*_{\text{sing}}$. The derivative $dI_{rk}$ is given by*

$$dI_{rk}(A) = (z-1)z^{-k} \int_{|\eta|=1} \text{tr}[adj((z-1)A(z)-\eta)] \frac{d\eta}{\eta^{n-r+1}} \in \underline{g}\ , \tag{2.81}$$

*where $adj(\cdot)$ denotes the classical adjoint matrix.*

**Proof:** The functions are clearly smooth (with respect to any reasonable topology on $\underline{\tilde{g}}^*_{\text{sing}}$) and the proof of (2.81) is a direct computation. By the results of $R$-matrix theory (see Appendix), to prove commutivity it is enough to show that the $I_{rk}$'s are invariant under the Ad$^*$ action of $G$,

$$I_{rk}(\text{Ad}^*_g A) = I_{rk}(A) \tag{2.82}$$

for all $g \in G$, where

$$(\text{Ad}^*_g A)(z) = g^{-1}(z)\,A(z)\,g(z) \in \underline{g}^*_{\text{sing}}\ . \tag{2.83}$$

But this follows trivially from the invariance of the determinant. □

Notice in particular from (2.81),

$$dI_{rk}(A)(z=1) = 0\ . \tag{2.84}$$

By general $R$-matrix theory, the $I_{rk}$'s generate flows of the form

$$\dot{A} = [\pi_- dI_{rk}(A), A] \tag{2.85}$$

on generic coadjoint orbits $O_A = O_{A_p/(z-1)+A_{-1}z^{-1}+A_0+A_1 z} \subset \underline{\tilde{g}}^*_{\text{sing}}$. Thus

$$\dot{A}_p = [(\pi_- dI_{rk}(A))(z),\ (z-1)A]\big|_{z=1} = 0\ , \tag{2.86}$$

by (2.84). Thus the entries of $A_p$ provide additional integrals for the $I_{rk}$ flows.

On generic orbits $O_A$ most of the $I_{rk}$'s degenerate: we compute the nontrivial intgrals that remain. For $(z-1)A(z) = A_p + (z-1)(A_{-1}z^{-1} + A_0 + A_1 z)$,

$$\det((z-1)A(z)-\eta) = \sum_{r=0}^{N} E_r(-z^{-1}A_{-1}+(A_p+A_{-1}-A_0)+z(A_0-A_1)+z^2A_1)\,\eta^{n-r}\ , \tag{2.87}$$

where $E_r(\cdot)$ is the elementary symmetric function of degree $r$ in the entries of $M$, $E_r(tM) = t^r E_r(M)$ for all $t \in R$. Thus

$$E_r(-z^{-1}A_{-1} + (A_p + A_{-1} - A_0) + z(A_0 - A_1) + z^2 A_1) = \sum_{k=-r}^{2r} I_{rk} z^k \ . \tag{2.88}$$

(Note that the notation is consistent: $\int_{|z|=1} \int_{|\eta|=1} I_{rk} z^k \eta^{n-r} \frac{\bar{d}\,\eta}{\eta^{n-r+1}} \frac{\bar{d}\,z}{z^{k+1}} = I_{rk}$.) It follows that the number of nontrivial integrals is given by

$$\sum_{r=1}^{N} (3r+1) = 3N(N+1)/2 + N$$

Simple computations show that the coadjoint invariants $\{\mathrm{tr}\ A_{-1}^k,\ \mathrm{tr}(A_p - A_0)A_{-1}^{k'}\}_{1 \le k,k' \le N}$ are equivalent to $\{I_{r,-r}, I_{r',-r'+1}\}_{1 \le r,r' \le N}$, and the coadjoint invariants $\{\mathrm{tr}\ A_1^k,\ \mathrm{tr}\ A_0 A_1^{k'}\}_{1 \le k,k' \le N}$ are equivalent to $\{I_{r,2r}, I_{r',2r'-1}\}_{1 \le r,r' \le N}$. We thus obtain

$$(3N(N+1)/2 + N) - 4N = 3N(N-1)/2$$

nontrivial integrals,

$$\{I_{rk}\}_{1 \le r \le N\ ,\ -r+2 \le k \le 2r-2} \tag{2.89}$$

on the orbit $O_A$. We construct the remaining $\frac{N(N-1)}{2}$ commuting integrals from the entries of $A_p$. By (2.86) it is enough to construct $N(N-1)/2$ combinations of the entries of $A_p$ that commute amongst themselves.

Note first that $(A_p)_{ij}$ can be rewritten in the form

$$(A_p)_{ij} = \Phi_{ij}(A) = \oint \left(\frac{z^4 - 1}{z^3}\right)\ \mathrm{tr}(A(z)E_{ji})\,\bar{d}\,z \tag{2.90}$$

where $E_{ji}$ is the elementary matrix $e_j e_i^T$. By differentiation

$$d\Phi_{ij}(A) = \left(\frac{z^4 - 1}{z^3}\right) E_{ji} \in \underline{g}\ , \tag{2.91}$$

and we find

$$R\, d\Phi_{ij}(A) = (z - z^{-1})^2 E_{ji}\ . \tag{2.92}$$

Substitution yields

$$\begin{aligned}\{\Phi_{ij}, \Phi_{rs}\}(A) &= (A, [d\Phi_{ij}(A)\ ,\ d\Phi_{rs}(A)]_R) \\ &= \oint \operatorname{tr}\left(A(z)\frac{(z^2-1)^2}{z^2}\frac{(z^4-1)}{z^3}[E_{ji}, E_{sr}]\right)\frac{đz}{z} \\ &= -\operatorname{tr} A_p[E_{ji}, E_{sr}] \\ &= -\delta_{is}(A_p)_{rj} + \delta_{rj}(A_p)_{is}\ .\end{aligned}$$

Thus

$$\{\Phi_{ij}, \Phi_{rs}\} = \delta_{rj}\Phi_{is} - \delta_{is}\Phi_{rj}\ . \tag{2.93}$$

Now consider $(A_p)_{ij} = \Phi_{ij}$ as a function on $g\ell(N, \boldsymbol{R})^*$ with the Lie-Poisson structure

$$\{F, G\}(A_p) = \operatorname{tr} A_p[\nabla F^T(A_p)\ ,\ \nabla G^T(A_p)]\ , \tag{2.94}$$

where $\nabla F(A_p) = (\dfrac{\partial F}{\partial (A_p)_{ij}})$. Then a straightforward computation shows that

$$\{\Phi_{ij}, \Phi_{rs}\}(A_p) = \delta_{is}\Phi_{rj} - \delta_{rj}\Phi_{is}\ , \tag{2.95}$$

and we see that, apart from an irrelevant minus sign, the problem reduces to the problem of computing Poisson brackets on $g\ell(N, \boldsymbol{R})^*$.

For $1 \leq r \leq N-1$, let

$$\lambda_{r1}, \ldots, \lambda_{rr} \tag{2.96}$$

denote the eigenvalues of the leading submatrix

$$\begin{pmatrix} (A_p)_{11} & \cdots & (A_p)_{1r} \\ \vdots & & \\ (A_p)_{r1} & \cdots & (A_p)_{rr} \end{pmatrix}.$$

(Note that the eigenvalues of the full matrix $A_p$ are already included in the set $\{I_{rk}\}$ in (2.89): set $z = 1$ in (2.87).) The $\lambda_{rk}$'s, $1 \leq k \leq r$, are invariant,

$$\lambda_{rk}(g^{-1}A_p g) = \lambda_{rk}(A_p)\ , \tag{2.97}$$

under conjugation by matrices of the form $g = \begin{pmatrix} g_r & 0 \\ 0 & I_{N-r} \end{pmatrix}$, where $g_r$ is $r \times r$. By differentiation, we find

$$\mathrm{tr}([\nabla\lambda_{rk}^T(A_p)\ ,\ A_p]X) = 0 \tag{2.98}$$

for all $X$ of the form $\begin{pmatrix} X_r & 0 \\ 0 & O_{N-r} \end{pmatrix}$. Suppose $r' \leq r$. Then by the computation of Thimm [Th],

$$\begin{aligned} \{\lambda_{rk}, \lambda_{r'k'}\}(A_p) &= \mathrm{tr}(A_p[\nabla\lambda_{rk}^T(A_p)\ ,\ \nabla\lambda_{r'k'}^T(A_p)]) \\ &= \mathrm{tr}([A_p, \nabla\lambda_{rk}^T(A_p)]\ \nabla\lambda_{r'k'}^T(A_p))\ . \end{aligned}$$

But as $r' \leq r$, $\nabla\lambda_{r'k'}^T(A_p)$ is clearly of the form $\begin{pmatrix} X_r & 0 \\ 0 & 0 \end{pmatrix}$) and so $\{\lambda_{rk}, \lambda_{r'k'}\}(A_p) = 0$ by (2.98). This provides the remaining $\sum_{r=1}^{N-1} r = N(N-1)/2$ commuting integrals.

We have proved the following result.

**Theorem 2.99.** *The functions $\{I_{rk}\}_{1\leq r\leq N,\,-r+2\leq k\leq 2r-2}$ and $\{\lambda_{rk}\}_{1\leq N-1,\,1\leq k\leq r}$ provide $2N^2 - 2N$ commuting integrals on an orbit $O_A$.* □

We leave it to the reader to verify that on a generic leaf $O_A$ of dimension $4N^2 - 4N$, the above integrals are independent on an open dense set. The values of the integrals for which the gradients become dependent correspond to separatrices for the associated flows.

### (d) Interpolating the Moser-Veselov algorithm

By analogy with the continuous-time Cholesky algorithm (see [DLT]), the Hamiltonian for the flow interpolating the Moser-Veselov iteration (1.22) with splitting (1.24) is given by

$$H(A) = \oint\ \mathrm{tr}\ (A(z)\log A(z) - A(z)\,)\frac{dz}{z}\ ,\ A \in \underline{\tilde{g}}^*_{\mathrm{sing}}\ . \tag{2.100}$$

As in the finite dimensional situation of [DLT], $H$ is clearly not globally defined on $\underline{\tilde{g}}^*_{\mathrm{sing}}$. But there is an additional difficulty in the infinite dimensional situation: the (formal) derivative of $H$ is given by

$$dH(A) = \log A = \log\left(\frac{A_p}{z-1} + A_{\mathrm{reg}}(z)\right)\ , \tag{2.101}$$

which can lie in $\underline{g}$ (see discussion following (2.42)) only if $A_p = 0$. Thus $H$ cannot be differentiable in an open set. However, $H$ is differentiable, with (directional) derivative

(2.101), at elements $A$ in the open subset of the finite codimensional submanifold $\underline{\tilde{g}}^*_{\text{reg}} \cap \{A(z) : A(1) = A_{\text{reg}}(1) = \text{diagonal}\}$ of $\underline{\tilde{g}}^*_{\text{sing}}$. In particular this is true of $A = A_{\text{reg}}(z)$ is positive definite on $S^1$ with $A(1)$ diagonal. As we will see, this is enough for our purposes.

In order to solve the interpolating flow by a Symes-type factorization, we must determine the map $\Theta$ (see Appendix) in our case $\underline{\tilde{g}}$ with $R$-matrix given by(1.37), (1.33) and (1.34). Simple computations show that the ideals $K_\pm$ are given by $\underline{g}_\pm \cap \{X(\lambda = \infty) = 0\}$ respectively, so that $\underline{g}_\pm / K_\pm$ is isomorphic to the diagonal matrices. Also $\theta(\delta) = -\delta$ for any diagonal matrix $\delta$, and condition (A.28) becomes

$$g_+(t, \lambda = \infty) = g_-(t, \lambda = \infty) \ . \tag{2.102}$$

We must also consider matrix factorization problems of the following form: suppose that $X(\lambda)$ is a self-adjoint element in $\underline{\tilde{g}}$,

$$X(\lambda) = X(\lambda)^* \ , \qquad \lambda \in i\boldsymbol{R} \ . \tag{2.103}$$

Then for any real $t$, $e^{(t/2)X(\infty)} \, e^{tX(\lambda)} \, e^{-(t/2)X(\infty)}$ is a smooth, positive definite matrix-valued function on $\dot{\Sigma}$, with limit

$$e^{-(t/2)X(\infty)} \, e^{tX(\lambda)} \, e^{-(t/2)X(\infty)} \to I \tag{2.104}$$

as $\lambda \to \infty$. By standard matrix factorization theory (see, for example, [G-K]), there exists a (unique) factorization

$$e^{-(t/2)X(\infty)} \, e^{tX(\lambda)} \, e^{-(t/2)X(\infty)} = \hat{g}_+(t, \lambda) \, \hat{g}_-(t, \lambda) \tag{2.105}$$

with $\hat{g}_\pm(t, \cdot)$ analytic in $\{\text{Re } \lambda > 0\}$, $\{\text{Re } \lambda < 0\}$ respectively, continuous and invertible in $\{\text{Re } \lambda \geq 0\}$, $\{\text{Re } \lambda \leq 0\}$ respectively, and satisfying

$$\hat{g}_\pm(t, \infty) = I \ . \tag{2.106}$$

Thus

$$\begin{aligned} g_+(t, \lambda) &\equiv e^{(t/2)X(\infty)} \, \hat{g}_+(t, \lambda) \\ g_-(t, \lambda) &\equiv \hat{g}_-(t, \lambda) \, e^{(t/2)X(\infty)} \end{aligned} \tag{2.107}$$

provides the (unique) factorization of $e^{tX(\lambda)}$,

$$e^{tX(\lambda)} = g_+(t,\lambda)\, g_-(t,\lambda)\ , \qquad g_\pm \in G\ , \tag{2.108}$$

with $g_\pm(t,\cdot)$ analytic in $\{\mathrm{Re}\ \lambda > 0\}$, $\{\mathrm{Re}\ \lambda < 0\}$ respectively, continuous and invertible in $\{\mathrm{Re}\ \lambda \geq 0\}$, $\{\mathrm{Re}\ \lambda \leq 0\}$ respectively, and satisfying (2.102). (Note that the requirement $g_\pm \in G$ implies, in particular, that $g_\pm(t,\infty) > 0$, which is needed for uniqueness.)

Now suppose that $(X_{-1}, X_0)$ is a pair in $Q^{2n} = O(N) \times O(N)$ for which the associated matrix $\omega_0 = X_0^T X_{-1}$ has spectrum $S_\pm$ satisfying (1.24). Then the associated loop

$$\widetilde{M}_0(\lambda) = \frac{1 - \lambda M_0 - \lambda^2 J^2}{1 - \lambda^2} = \frac{(\omega_0^T + \lambda J)(\omega_0 - \lambda J)}{1 - \lambda^2}$$

is invertible on $\dot{\Sigma}$. Moreover, as the loop is clearly self-adjoint on $\dot{\Sigma}$, with limit $J^2 > 0$ at infinity, the loop is necessarily (strictly) positive definite. It follows that

$$X_0(\lambda) \equiv \log\left(\widetilde{M}_0(\lambda)\right) \tag{2.109}$$

(take the principal branch of the logarithm) is self-adjoint on $\dot{\Sigma}$, and as $X_0(\infty) = log J^2$ is diagonal, $X_0(\lambda) \in \underline{\tilde{g}}$. Thus $e^{tX_0(\lambda)} \in \widetilde{G}$, and from the previous discussion, there exists a factorization

$$e^{tX_0(\lambda)} = g_+(t,\lambda)\, g_-(t,\lambda) \tag{2.110}$$

of type (2.103) satisfying (2.102). By uniqueness and self-adjointness we must have

$$g_+(t,\lambda) = g_-(t,\bar{\lambda})^*\ , \tag{2.111}$$

and by reality, $\overline{\widetilde{M}_0(\lambda)} = \widetilde{M}_0(\overline{\lambda})$,

$$g_\pm(t,\lambda) = \overline{g_\pm(t,\bar{\lambda})}\ . \tag{2.112}$$

By the above calculations and by $R$-matrix theory, we have proved most of the following basic result:

**Theorem 2.113.** *Let*

$$e^{t\log(\widetilde{M}_0(\lambda))} = g_+(t,\lambda)\,g_-(t,\lambda)$$

*be the factorization* (2.110) *above. Then*

$$\begin{aligned}\widetilde{M}(t,\lambda) &\equiv g_+(t,\lambda)^{-1}\widetilde{M}_0(\lambda)\,g_+(t,\lambda)\\ &= g_-(t,\lambda)\,\widetilde{M}_0(\lambda)\,g_-(t,\lambda)^{-1}\end{aligned} \tag{2.114}$$

*solves the flow*

$$\begin{aligned}\frac{d\widetilde{M}(t,\lambda)}{dt} &= \left[(\pi_-\log\widetilde{M}(t,\cdot)\,)(\lambda)\ ,\ \widetilde{M}(t,\lambda)\right]\ ,\\ \widetilde{M}(0,\lambda) &= \widetilde{M}_0(\lambda)\end{aligned} \tag{2.115}$$

*generated by the* Ad$^*$*-invariant Hamiltonian*

$$H(A) = \lim_{r\uparrow\infty}\int_{-ir}^{ir}\operatorname{tr}\big(A(\lambda)\log A(\lambda) - A(\lambda)\big)\,\frac{2đ\lambda}{1-\lambda^2} \tag{2.116}$$

*on* $\tilde{\underline{g}}^*_{\text{sing}}$*, and at integer times interpolates the Moser-Veselov algorithm*

$$\widetilde{M}(k,\lambda) = M_k(\lambda)\,/\,(1-\lambda^2)\ , \qquad k\in\mathbb{Z}\ . \tag{2.117}$$

*For all times* $t$, $\widetilde{M}(t,\lambda)$ *has the form*

$$\widetilde{M}(t,\lambda) = (I - \lambda M(t) - \lambda^2 J^2)\,/\,(1-\lambda^2)\ , \tag{2.118}$$

*where* $M(t)$ *is real and skew, and* $M(k) = M_k$ *for all* $k\in\mathbb{Z}$.

**Proof:** Equation (2.115) follows from $R$-matrix theory and can be verified directly. One obtains the formula

$$\log\widetilde{M}(t,\lambda) = g_+(t,\lambda)^{-1}\,\frac{dg_+(t,\lambda)}{dt} + \frac{dg_-(t,\lambda)}{dt}\,g_-(t,\lambda)^{-1}\ ,$$

and one needs

$$\frac{dg_-(t,\lambda)}{dt}g_-(t,\lambda)^{-1} = (\pi_-\log\widetilde{M}(t,\cdot)\,)(\lambda)\ . \tag{2.119}$$

But if one writes $(dg_-(t,\lambda)/dt)\, g_-(t,\lambda)^{-1} = \gamma_- + \gamma_0$ in the $z$-notation (2.7), then

$$\pi_-(dg_-(t,\cdot)/dt)\, g_-(t,\cdot)^{-1} = \gamma_- + (\gamma_0 - \gamma_-(1))/2 . \tag{2.120}$$

On the other hand

$$\begin{aligned}\pi_- g_+(t,\cdot)^{-1}\Big(\frac{dg_+(t,\cdot)}{dt}\Big) = \text{constant } &= \frac{1}{2} g_+(t,z=1)^{-1}\, \frac{dg_+(t,z+1)}{dt} \\ &= \frac{1}{2} g_-(t,z=1)^{-1}\, \frac{d}{dt} g_-(t,z=1)\,, \quad \text{by (2.102)}, \\ &= \frac{1}{2}\Big(\frac{d}{dt} g_-(t,z=1)\Big) g_-(t,z=1)^{-1}\,, \quad \text{by diagonality}, \\ &= \frac{1}{2}(\gamma_-(1) + \gamma_0)\,. \end{aligned} \tag{2.121}$$

Adding (2.120) and (2.121) we obtain (2.119).

To prove (2.118) note first from (2.114) that $(1-\lambda^2)\widetilde{M}(t,\lambda)$ has an analytic continuation to $\operatorname{Re}\lambda > 0$ and to $\operatorname{Re}\lambda < 0$, and grows at most quadratically as $\lambda \to \infty$. By Liouville,

$$\widetilde{M}(t,\lambda) = \frac{B_0 + \lambda B_1 + \lambda^2 B_2}{1-\lambda^2}$$

for suitable matrices $B_0, B_1$ and $B_2$. Letting $\lambda \to \infty$ in (2.114), and using the diagonality of $g_+(t,\lambda=\infty)$, we see that $B_2 = -J^2$. Letting $\lambda \to 0$, we find

$$B_0 = \widetilde{M}(t,0) = g_+(t,0)^{-1} I\, g_+(t,0) = I\,.$$

From (2.111), for $\lambda \in i\boldsymbol{R}$,

$$\begin{aligned}\widetilde{M}(t,\lambda)^* &= g_+(t,\lambda)^*\Big(\frac{I + \lambda M_0^T - \lambda^2 J^2}{1-\lambda^2}\Big)(g_+(t,\lambda)^{-1})^* \\ &= g_-(t,\lambda)\Big(\frac{I - \lambda M_0 - \lambda^2 J^2}{1-\lambda^2}\Big)(g_-(t,\lambda))^{1} \quad (M_0 \text{ is skew}) \\ &= \widetilde{M}(t,\lambda)\,,\end{aligned}$$

and hence $B_1 = -B_1^*$. But using (2.112), we find $M(t,\lambda) = \overline{M(t,\bar{\lambda})}$, and so $B_1$ is real. It follows that

$$\widetilde{M}(t,\lambda) = \frac{I - \lambda M(t) - \lambda^2 J^2}{1-\lambda^2}\,,$$

where $M(t)$ is real and skew.

Finally, to prove (2.117), it is sufficient,by the group property for flows, to verify (2.117) for $k = 1$. But for $k = 1$,

$$
\begin{aligned}
e^{\log(\widetilde{M}_0(\lambda))} &= \frac{I - \lambda M_0 - \lambda^2 J^2}{1-\lambda^2} \\
&= \Big(\frac{\omega_0^T + \lambda J}{1+\lambda}\Big)\Big(\frac{\omega_0 - \lambda J}{1-\lambda}\Big), \quad \text{by (1.20).}
\end{aligned}
$$

But these factors satisfy (1.29), (1.30) and so

$$
\begin{aligned}
g_-(1,\lambda) &= (\omega_0 - \lambda J)/(1-\lambda) \\
g_+(1,\lambda) &= (\omega_0^T + \lambda J)/(1+\lambda)
\end{aligned}
\tag{2.122}
$$

by uniqueness. Thus

$$
\begin{aligned}
\frac{I - \lambda M(1) - \lambda^2 J^2}{1-\lambda^2} = \widetilde{M}(1,\lambda) &= g_+(1,\lambda)^{-1}\Big(\frac{I - \lambda M_0 - \lambda^2 J^2}{1-\lambda^2}\Big) g_+(1,\lambda) \\
&= \Big(\frac{\omega_0^T + \lambda J}{1+\lambda}\Big)^{-1}\Big(\frac{\omega_0^T + \lambda J}{1+\lambda}\Big)\Big(\frac{\omega_0 - \lambda J}{1-\lambda}\Big)\Big(\frac{\omega_0^T + \lambda J}{1+\lambda}\Big) \\
&= \frac{M_1(\lambda)}{1-\lambda^2} \\
&= \frac{I - \lambda M_1 - \lambda^2 J^2}{1-\lambda^2},
\end{aligned}
$$

which completes the proof. □

The Hamiltonian $H$ generates (2.115) on the coadjoint orbit $O_{A^0}$ where

$$
\begin{aligned}
A^0 &= (1 - \lambda M_0 - \lambda^2 J^2)/(1-\lambda^2) \\
&= \frac{A^0_{-1}}{z} + A^0_0 + A^0_1 z,
\end{aligned}
$$

and

$$
A^0_1 = (J^2 - M_0 - 1)/4,\ A^0_0 = (1 + J^2)/2 \text{ and } A^0_{-1} = (J^2 + M_0 + 1)/4 \tag{2.123}
$$

(see Section 2(b)). By the results of the previous subsection $O_{A^0}$ is generic and of dimension $4N^2 - 4N$ provided $A^0_1$ (and hence $A^0_{-1} = (A^0_1)^T$) is invertible and has distinct eigenvalues (given $J$, this is clearly true for a dense open set of $M_0$'s). The $I_{rk}$'s and the $\lambda_{rk}$'s of

Theorem 2.49 provide $2N^2 - 2N$ commuting integrals for $H$. Indeed, $\{I_{rk}, H\}_{\underline{g}^*} = 0$ by $\mathrm{Ad}^*$-invariance and $R$-matrix theory: alternatively from (2.114),

$$\begin{aligned} I_{rk}(\widetilde{M}(t,\cdot)) &= I_{rk}(g_+(t,0)^{-1}\, \widetilde{M}(t,\cdot)\, g_+(t,0)) \\ &= I_{rk}(\widetilde{M}(0,\cdot)) \ . \end{aligned}$$

On the other hand, the conservation of the $\lambda_{rk}$'s is trivial. The integrals $I_{rk}$, $\lambda_{rk}$, however, are clearly dependent on the invariant set

$$\begin{aligned} \mathcal{M}_{A^0} = \{A \in O_{A^0} : I_{rk}(A) = I_{rk}(A^0)\ ,\ 1 \le r \le N\ ,\ -r+2 \le k \le 2r-2\ , \\ \lambda_{rk}(A) = \lambda_{rk}(A^0) = 0\ ,\ 1 \le r \le N-1,\ 1 \le k \le r\} \end{aligned}$$

and so the $H$-flow with the relevant initial conditions always lies on a separatrix of the integrable scheme. The flow on the separatrix can be analyzed by noting that $H$ induces a flow $t \to M(t)$ on the reduced space $o(N)^*$. As we now show, and as anticipated from [MV], this flow is completely integrable in the classical sense.

The equation of motion for $M(t)$ is obtained by differentiating (2.115) with respect to $\lambda$ and setting $\lambda = 0$. We find

$$\begin{aligned} \frac{dM(t)}{dt} &= \left[(\pi_- \log \widetilde{M}(t,\cdot))(0)\ ,\ M(t)\right]\ , \\ M(0) &= M_0\ . \end{aligned} \tag{2.124}$$

From (2.114) we have

$$I - \lambda M(t) - \lambda^2 J^2 = I - \lambda\, g_+(t,\lambda)^{-1}\left[M_0 + \lambda J^2\right] g_+(t,\lambda)$$

and letting $\lambda \to 0$,

$$M(t) = g_+(t,0)^{-1} M_0\, g_+(t,0)\ . \tag{2.125}$$

But more is true: $M(t) = -M(t)^T$ is orthogonally equivalent to $M_0$, as it should be. Indeed if

$$g_+(t,\lambda) = |g_+(t,\lambda|\, Q_+(t,\lambda)\ , \quad \lambda \in i\boldsymbol{R} \tag{2.126}$$

is the polar decomposition of $g_+(t,\lambda)$, then using (2.111) in (2.110), we obtain

$$|g_+(t,\lambda)| = e^{(t/2)\log(\widetilde{M}_0(\lambda))} \tag{2.127}$$

In particular $|g_+(t,0)| = I$ and so $g_+(t,0) = Q_+(t,0)$. But by (2.112), $g_+(t,0)$ is real. Hence $Q_+(t,0)$ is real and orthogonal, and

$$M(t) = Q_+(t,0)^T M_0 \, Q_+(t,0) \, . \tag{2.128}$$

Of course for $t=1$, $Q_+(t,0) = g_+(t,0) = \omega_0^T$ from (2.122), as it should (see (1.12)).

We identify $o(N)^*$ with $o(N)$ through the standard, nondegenerate pairing $(B_1, B_2) = \text{tr } B_1 B_2$. For smooth functions $F_i$, $1 \le i \le 2$, from $G\ell(n, \mathbf{R})$ to $\mathbf{C}$, the Lie-Poisson bracket on $o(N)^*$ is given by

$$\{F_1, F_2\}_{o(N)^*}(M) = \text{ tr } M\big[\pi_0(\nabla F_1(m)) \, , \, \pi_0(\nabla F_2(M))\big] \tag{2.129}$$

where $\nabla F_i(M) = (\dfrac{\partial F_i(M)}{\partial M_{jk}})$ and $\pi_0 A = (A - A^T)/2$. Now as noted above, $I - \lambda M_0 - \lambda^2 J^2$ is positive definite on $i\mathbf{R}$, and by (2.114), the same is true for $I - \lambda M(t) - \lambda^2 J^2$. On the other hand, for a general matrix $B \in o(N)$,

$$\begin{gathered} I - \lambda B - \lambda^2 J^2 > 0 \qquad \text{on} \quad i\mathbf{R} \\ \Leftrightarrow \\ \|u\|^2 - \gamma(\pi, iBu) + \gamma^2 \|Ju\|^2 > 0 \quad \text{for all} \quad \gamma \in \mathbf{R} \\ \Leftrightarrow \\ |||B||| \equiv \sup_{u \neq 0} \frac{|(\bar{u}, iBu)|}{\|Ju\| \, |||u|||} < 2 \, . \end{gathered} \tag{2.130}$$

Thus the dynamics $t \to M(t)$ takes place on the Poisson manifold

$$o_2(N)^* \equiv \{M \in o(N)^* : |||M||| < 2\} \, . \tag{2.131}$$

To show that the flow $t \to M(t)$ is Hamiltonian on $o_2(N)^*$, we first consider a slightly more general situation. Let $F$ be a real analytic function on $(0, \infty)$. Then the preceding calculations show that for initial data

$$A^0(\lambda) = \frac{1 - \lambda M_0 - \lambda^2 J^2}{1 - \lambda^2} \, , \qquad M_0 \in o_2(N)^* \, ,$$

the Hamiltonian

$$H_F = H_F(A) = \int_{-i\infty}^{i\infty} \operatorname{tr} F(A) \frac{2\bar{d}\lambda}{1-\lambda^2} \tag{2.132}$$

generates a flow

$$\frac{dA_F(t,\lambda)}{dt} = \left[(\pi_- F'(A_F(t,\cdot)))(\lambda) \, , \, A_F(t,\lambda)\right] \tag{2.133}$$

where

$$A_F(t,\lambda) = \frac{I - \lambda M_F(t) - \lambda^2 J^2}{1-\lambda^2} \, , \quad M_F(t) \in o_2(N)^* \, . \tag{2.134}$$

Differentiation with respect to $\lambda$ at $\lambda = 0$ again gives

$$\frac{dM_F(t)}{dt} = \left[(\pi_- F'(A_F(t,\cdot)))(0) \, , \, M_F(t)\right] . \tag{2.135}$$

Let $\int' = \lim_{r\uparrow\infty} \int_{-ir}^{ir}{}'$ denote integration along the indented contour

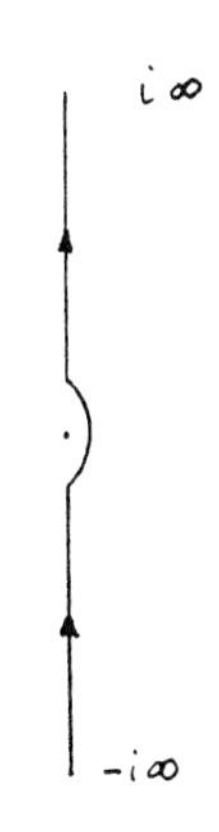


Figure 2.1

As $F$ is real analytic, we find

$$(\pi_- F'(A_F))(0) = \int' F'(A_F(t,\lambda)) \frac{\bar{d}\lambda}{\lambda} . \tag{2.136}$$

On the other hand a Hamiltonian $\widehat{H}$ on $o_2(N)^*$ generates the flow

$$\frac{dM}{dt} = \left[\pi_0 \nabla \widehat{H}(M), M\right] . \tag{2.137}$$

For $M \in o_2(N)^*$ set

$$\widehat{H} = \widehat{H}_F(M) = \int' \text{tr}\ \left[F(\widetilde{M}(\lambda)) - F(J^2)\right] \frac{đ\lambda}{\lambda^2(1-\lambda^2)^{-1}}\ . \tag{2.138}$$

A direct computation shows that

$$\nabla \widehat{H}_F(M) = -\int' \left(F'(\widetilde{M}(\lambda))\right)^T \frac{đ\lambda}{\lambda}\ . \tag{2.139}$$

Substitution of (2.139) in (2.136) yields (2.135). In particular for $F(x) = x\log x - x$, *we see that $\widehat{H}_F$ is the Hamiltonian for the flow induced on $o_2(N)^*$ by the interpolating flow* (2.115). In particular the time-one map $M_0 \to M(1) = M_1$ is Poisson on $o_2(N)^*$, as discovered by Moser and Veselov [MV].

The above computations can also clearly be used to show that

$$\widehat{I}_{rk}(M) = I_{rk}(\widetilde{M}(\lambda))$$

forms a commuting set of integrals on $o_2(N)^*$ for $\widehat{H}_F$. The dimension of the generic symplectic leaves of the Lie-Poisson structure on $o_2(N)^*$ is $\frac{1}{2}N(N-1) - [N/2]$, and one verifies by a straightforward combinatorial argument that, in the generic case, precisely $\frac{1}{2}(\frac{1}{2}N(N-1) - [N/2])$ of the $I_{rk}$'s remain independent on the leaf. These integrals are of course equivalent to the eigenvalues of the operator $M + \lambda J^2$, which were used by Manakov [Man] to integrate the continuum Euler-Arnold equations (see *application* below).

We have proved the following result, which is the final theorem of this section.

**Theorem 2.140.** *The interpolating flow* (2.115) *on $O_{(I-\lambda M-\lambda^2 J^2)/(1-\lambda^2)}$, induces a completely integrable Hamiltonian flow $t \to M(t)$ on $o_2(N)^*$. The induced flow has solution*

$$M(t) = g_+(t,0)^{-1}\, M_0\, g_+(t,0)\ , \tag{2.141}$$

*where $g_+(t,0)$ is orthogonal and is the evaluation at $\lambda = 0$ of the solution $g_+(t,\lambda)$ of the Riemann-Hilbert factorization problem* (2.110). □

As an *application* of the methods of the chapter we now show how to use the interpolating flow to resolve a question raised by Moser and Veselov in [MV].

The continuum Euler-Arnold equations for the motion of an $N$-dimensional rigid body are given by

$$\frac{dM}{dt} = [M, \Omega] \ , \tag{2.142}$$

where $\Omega \in o(N)$ is defined through

$$M = J\Omega + \Omega J \ . \tag{2.143}$$

These equations are the continuum limit of the discrete Euler-Arnold equation (1.12) (see [MV]) and, as noted above, they have the same integrals. Question: is equation (1.12) the time-$\tau$ map for (2.142) for some $\tau > 0$?

The answer is "no" by the following argument. The interpolating flow for (1.12) is

$$\frac{dM}{dt} = \left[(\pi_- \log \widetilde{M}(t, \cdot\,))(0)\,,\ M\right] \ . \tag{2.124}$$

Evaluating $(\pi_- \log \widetilde{M}(t, \cdot\,))(0)$ for small $M$, we find after a straightforward calculation that

$$\left[(\pi_- \log \widetilde{M}(t, \cdot)(0), M\right] = -[\Omega, M] + K(M) + O(M^6) \ , \tag{2.144}$$

where

$$K(M) = -\lim_{r\to\infty} \int_{-ir}^{ir} \Big( \int_C \Big[\frac{1}{s - \frac{1-\lambda^2 J^2}{1-\lambda^2}} (M \frac{1}{s - \frac{1-\lambda^2 J^2}{1-\lambda^2}})^3, M\Big] \log s \frac{ds}{2\pi i}\Big) \frac{\lambda^2}{(1-\lambda^2)^3} \frac{d\lambda}{2\pi i} \ , \tag{2.145}$$

and so (2.124) becomes

$$\frac{dM}{dt} = [M, \Omega] + K(M) + O(M^6) \ ,$$

which agrees with (2.142) to third order. However a simple computation shows that $K(M)$ is not identically zero for all $J$ and $M$, and hence (2.142) cannot interpolate (1.12) for any $\tau > 0$.

As a final remark we note that in the continuum limit, $M = \epsilon m(\epsilon t)$, $\epsilon \downarrow 0$, equation (2.124) reduces in scaled time $T = \epsilon t$ to the equation

$$\frac{d}{dT}(m + \mu J^2) = \left[m + \mu J^2,\ \Omega + \mu J\right] \ ,$$

$$m = J\Omega + \Omega J \ ,$$

which is Manakov's Lax-pair form for the Euler-Arnold equations (see [Man]).

## 3. The discrete Euler-Arnold equation (II).

In this section we construct an interpolating flow for the discrete Euler-Arnold equation in the general case (1.19). Our result is Theorem 3.58 below.

For ease of presentation, we first consider the factorization (1.19) under the additional simplifying assumption, removed eventually in Remark 3.59 below, that if

$$\left.\begin{array}{l} \lambda, \lambda' \in S\,, \\ \text{and} \\ \operatorname{Re}\lambda = \operatorname{Re}\lambda'\,, \\ \text{then} \\ \lambda \text{ and } \lambda' \text{ both belong to } S_+, \text{ or both belong to } S_-\,. \end{array}\right\} \tag{3.1}$$

Under this assumption, a factorization of the form

$$\frac{I - \lambda M - \lambda^2 J^2}{1 - \lambda^2} = \left(\frac{\omega^T + \lambda J}{1 + \lambda}\right)\left(\frac{\omega - \lambda J}{1 - \lambda}\right),$$

where $S_\pm$ satisfy (1.19), clearly gives rise to factors $\dfrac{\omega^T + \lambda J}{1 + \lambda}$ and $\dfrac{\omega - \lambda J}{1 - \lambda}$ which are analytic and invertible in strips parallel to the imaginary axis. For this reason we consider a skeleton

$$\Sigma = \bigcup_{j=-m}^{m} \Sigma_j \tag{3.2}$$

consisting of $2m+1$ vertical lines $\Sigma_j = x_j + iR$, positioned, without loss of generality, such that

$$\begin{gathered} x_{-m} < x_{-m+1} < \cdots < x_{-1} < x_0 = 0 < x_1 < \cdots < x_m\,, \\ x_{-j} = -x_j\,. \end{gathered} \tag{3.3}$$

dividing $\mathbb{C}$ into $2m+2$ strips $\Omega_{-m}, \ldots, \Omega_0, \Omega_1, \ldots, \Omega_{m+1}$. We attach signatures to the $\Omega_j$'s as follows,

$$\operatorname{sgn}\Omega_j = (-1)^{m+1-j}\,, \qquad -m \le j \le m+1 \tag{3.4}$$

(note that we always have $\operatorname{sgn}\Omega_{m+1} = 1$, $\operatorname{sgn}\Omega_{-m} = -1$) and define a decomposition of

$\mathbb{C} \setminus \Sigma$ by

$$\begin{aligned} \Omega_+ &\equiv \bigcup_{\operatorname{sgn}\,\Omega_j > 0} \Omega_j \\ \Omega_- &\equiv \bigcup_{\operatorname{sgn}\,\Omega_j < 0} \Omega_j \end{aligned}, \tag{3.5}$$

$\Omega_+ \cap \Omega_- = \emptyset$, $\Omega_+ \cup \Omega_- = \mathbb{C} \setminus \Sigma$. Thus for $m = 1$, say, we have

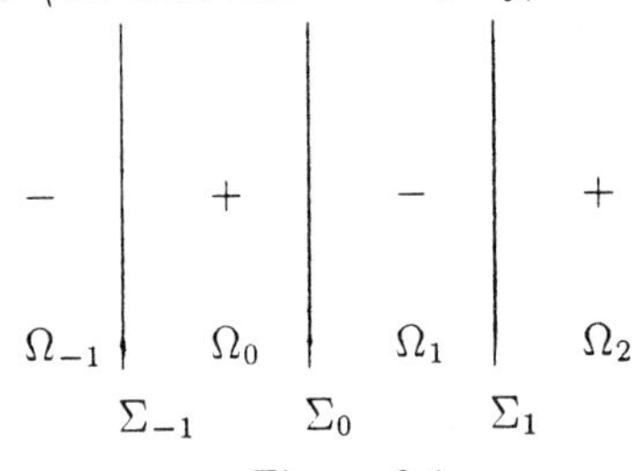

Figure 3.1

and for $m = 2$,

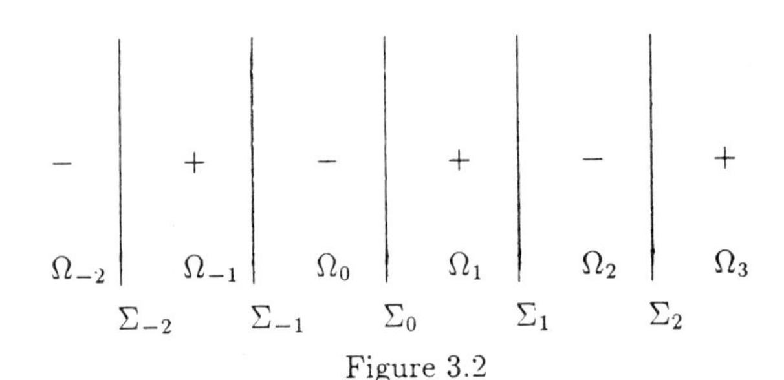

Figure 3.2

Let $G$ be the group of all smooth functions $g : \Sigma \to G\ell(N, \mathbb{C})$ such that

(i) $g(\lambda)$ and all its derivatives have limits at $x_k \pm i\infty$ for $-m \leq k \leq m$,

(ii) $g(x_k \pm i\infty)$ is diagonal with positive diagonal entries at $x_k + i\infty$, $-m \leq k \leq m$,

(iii) ("zero winding condition")

$$\begin{aligned} &g(x_m + i\infty)\, g(x_{m-1} + i\infty)^{-1}\, g(x_{m-2} + i\infty)\, g(x_{m-3} + i\infty)^{-1} \\ &\qquad \cdots g(x_{-m+1} + i\infty)^{-1}\, g(x_{-m} + i\infty) \\ &= \\ &g(x_m - i\infty)\, g(x_{m-1} - i\infty)^{-1}\, g(x_{m-2} - i\infty)\, g(x_{m-3} - i\infty)^{-1} \\ &\qquad \cdots g(x_{-m+1} - i\infty)^{-1}\, g(x_{-m} - i\infty)\,, \end{aligned}$$

and the group operation is pointwise multiplication on $\Sigma$. Note that the case $m = 0$ reduces (essentially) to the group $G$ of Section 2.

The Lie-algebra $\underline{g}$ associated to $G$ is the set of all smooth functions $X : \Sigma \to g\ell(N, \mathbb{C})$ such that

(i)′ $X(\lambda)$ and all its derivatives have limits at $x_k \pm i\infty$ for $-m \le k \le m$,

(ii)′ $X(x_k \pm i\infty)$ is diagonal with real entries at $x_k + i\infty$, $-m \le k \le m$,

(iii)′ ("zero winding condition")

$$\sum_{k=-m}^{m} (-1)^{m-k} X(x_k + i\infty) = \sum_{k=-m}^{m} (-1)^{m-k} X(x_k - i\infty) ,$$

and the bracket operation is taken pointwise on $\Sigma$.

The skeleton $\Sigma$ has two natural orientations associated with $\Omega_+$ and $\Omega_-$ respectively. The *$(\pm)$-orientations* of $\Sigma$ are specified by requiring that $\Omega_\pm$ always lies to the *left* as $\Sigma$ is traversed, respectively. Thus, for example, in the case that $m = 1$, we have

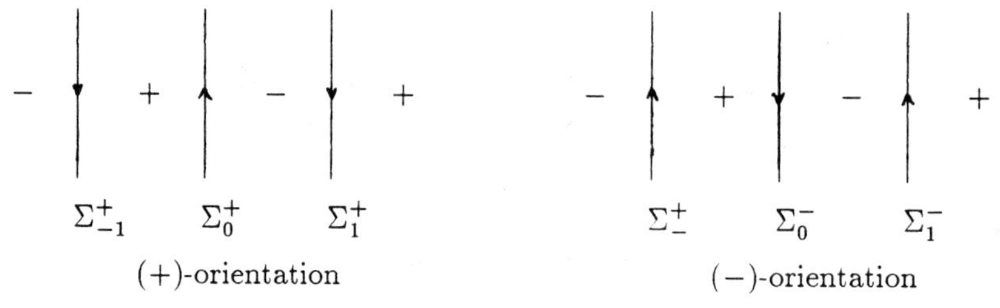


Figure 3.3

We denote $\Sigma$ with the $(\pm)$-orientation as $\Sigma^\pm$ respectively, and

$$\Sigma^\pm = \bigcup_{j=-\infty}^{\infty} \Sigma^\pm_j . \tag{3.6}$$

in a natural notation.

We will use an auxiliary function $\gamma(\lambda)$ with the following properties:

(i)″ $\gamma(\lambda)$ is analytic in a neighborhood of $\overline{\left(\bigcup_{j=-m+1}^{m} \Omega_j\right)}$

(ii)″ $\gamma(x + i\infty) = 0$, $\gamma(x - i\infty) = 1$ for $x$ in a neighborhood of $(x_{-m}, x_m)$.

We may take, for example,

$$\begin{aligned}\gamma(\lambda) &= (\log(\lambda - b) - \log(\lambda - a))/2\pi i \\ &= \frac{1}{2\pi i}\log\left|\frac{\lambda - b}{\lambda - a}\right| + \frac{\arg(\lambda - b) - \arg(\lambda - a)}{2\pi}\end{aligned} \tag{3.7}$$

where $0 < \arg(\lambda - b) < 2\pi$, $-\pi < \arg(\lambda - a) < \pi$ and $a, b$ are any real numbers for which $a < a_{-m} < \cdots < x_m < b$, but any $\gamma$ with properties (i)″ and (ii)″ would do.

For $X \in \underline{g}$ and $\lambda \in \Sigma$, set

$$\begin{aligned}(\pi_\pm X)(\lambda) &\equiv \lim_{\substack{\lambda'' \to \lambda \\ \lambda'' \in \Omega_\pm}} \int_{\Sigma\pm} X(\lambda') \frac{đ\lambda'}{\lambda' - \lambda''} \\ &\equiv \lim_{\substack{\lambda'' \to \lambda \\ \lambda'' \in \Omega_\pm}} \lim_{R \to \infty} \sum_{i=-m}^{m} \int_{\substack{\Sigma_j^\pm \\ |\mathrm{Im}\ \lambda'| \leq R}} X(\lambda') \frac{đ\lambda'}{\lambda' - \lambda''} \,.\end{aligned}$$

Clearly $(\pi_\pm X)(\lambda)$ have analytic continuations

$$\int_{\Sigma\pm} X(\lambda') \frac{đ\lambda'}{\lambda' - \lambda}$$

to $\Omega_\pm$ respectively.

To show that the above principal values exist as $R \to \infty$, we rewrite

$$\begin{aligned}&\int_{\substack{\Sigma\pm \\ |\mathrm{Im}\ \lambda'| \leq R}} X(\lambda') \frac{đ\lambda'}{\lambda' - \lambda''} \\ &= \sum_{j=-m}^{m} \int_{\substack{\Sigma_j^\pm \\ |\mathrm{Im}\ \lambda'| \leq R}} [X(\lambda') - (X(x_j + i\infty) + \gamma(\lambda')(X(x_j - i\infty) - X(x_j + i\infty)))] \frac{đ\lambda'}{\lambda' - \lambda''} \\ &\quad + \sum_{j=-m}^{m} \int_{\substack{\Sigma_j^\pm \\ |\mathrm{Im}\ \lambda'| \leq R}} (X(x_j + i\infty) + \gamma(\lambda')(X(x_j - i\infty) - X(x_j + i\infty))) \frac{đ\lambda'}{\lambda' - \lambda''} \,.\end{aligned} \tag{3.8}$$

The first sum has a limit as $R \to \infty$ as all the integrands are in $L^1$. By Cauchy's Theorem, in the case $\pi_+ X$, the second sum takes the form

$$\int_{\substack{\Sigma^{\pm}_{-m} \\ |\mathrm{Im}\,\lambda'|\le R}} \sum_{j=-m}^{m} (-1)^{m-j}\big(X(x_j+i\infty)+\gamma(\lambda')\,(X(x_j-i\infty)-X(x_j+i\infty))\big)\frac{d\lambda'}{\lambda'-\lambda''}$$

$$+ \sum_{\substack{j=-m \\ \mathrm{Re}\,\lambda''<x_j}} \big(X(x_j+i\infty)+\gamma(\lambda'')(X(x_j-i\infty)-X(x_j+i\infty))\big)(-1)^{m-j+1}$$

$$= \frac{1}{2}\sum_{j=-m}^{m}(-1)^{m-j}X(x_j+i\infty) + \sum_{\substack{j=-m \\ \mathrm{Re}\,\lambda''<x_j}}^{m} \big(X(x_j+i\infty)+\gamma(\lambda'')(X(x_j-i\infty)$$

$$- X(x_j+i\infty))\big)(-1)^{m-j+1}, \tag{3.9$_+$}$$

by (iii)$'$, and in the case $\pi_- X$, the second sum becomes

$$\frac{1}{2}\sum_{j=-m}^{m}(-1)^{m-j}X(x_j+i\infty)$$

$$+ \sum_{\substack{j=-\infty \\ \mathrm{Re}\,\lambda''>x_j}}^{m} \big(X(x_j+i\infty)+\gamma(\lambda'')(X(x_j-i\infty)-X(x_j+i\infty))\big)(-1)^{m-j+1}. \tag{3.9$_-$}$$

Together with standard results for the Cauchy transform, the above results show that $(\pi_\pm X)(\lambda)$ exist.

The first sum in (3.8) makes zero contribution to $\pi_\pm X(\lambda)$ as $\lambda \to \infty$, $\lambda \in \Sigma$, and we evaluate $\pi_\pm X(x_j \pm i\infty)$ by letting $\lambda \to \infty$ in $(3.9)_+$ and $(3.9)_-$. By inspection we see that

(a) $\pi_\pm X(x_j \pm i\infty)$ are diagonal and $(\pi_\pm X)(x_j + i\infty)$ are real.

(b) if $\Omega_j \subset \Omega_+$, then

$$(\pi_+ X)(x_{j-1} \pm i\infty) = (\pi_+ X)(x_j \pm i\infty)\,. \tag{3.10$_+$}$$

and if $\Omega_j \subset \Omega_-$, then

$$(\pi_- X)(x_{j-1} \pm i\infty) = (\pi_- X)(x_j \pm i\infty)\,. \tag{3.10$_-$}$$

(c) in general $(\pi_\pm X)(x_j + i\infty) \neq (\pi_\pm X)(x_j - i\infty)$, but for $j = m, -m$,

$$(\pi_+ X)(x_m + i\infty) = (\pi_+ X)(x_m - i\infty) = (\pi_- X)(x_{-m} + i\infty) = (\pi_- X)(x_{-m} - i\infty) \tag{3.11}$$

with common value

$$\frac{1}{2}\sum_{j=-m}^{m}(-1)^{m-j}X(x_j+i\infty)=\frac{1}{2}\sum_{j=-m}^{m}(-1)^{m-j}X(x_j-i\infty)\ . \tag{3.12}$$

Formulae (3.11) and (3.12) give the appropriate generalization of the case when $m=0$ (cf. (1.36)). From $(3.10)_\pm$ and (3.11) we see that $\pi_\pm X$ satisfy the zero winding condition (iii)$'$ and *we conclude that* $\pi_\pm X$ *map* $\underline{g}$ *into* $\underline{g}$. The standard calculation shows that

$$\pi_+ + \pi_- = 1\ . \tag{3.13}$$

Furthermore, a similar calculation to that in Section 2 shows that

$$R=\pi_+ - \pi_- \tag{3.14}$$

solves the modified Yang-Baxter equation on $\underline{g}$, and hence $\tilde{\underline{g}}=(\underline{g},[\cdot,\cdot]_R)$ is a Lie-algebra. Again we define the subalgebras

$$\underline{g}_\pm = \text{Ran}\ \pi_\pm\ , \qquad \underline{g}_+ + \underline{g}_- = \underline{g}\ . \tag{3.15}$$

One finds that the ideals $K_\pm \subset \underline{g}_\pm$ (see Appendix) are given by

$$\begin{aligned} K_+ &= \underline{g}_+ \cap \{X : X(x_m+i\infty)=X(x_m-i\infty)=0\} \\ K_- &= \underline{g}_- \cap \{X : X(x_{-m}+i\infty)=X(x_{-m}-i\infty)=0\} \end{aligned}\ . \tag{3.16}$$

This leads to the fóllowing factorization problem.

Given $g\in G$, find $g_\pm \in e^{\underline{g}_\pm}\subset G$, analytic in $\Omega_\pm$ respectively, continuous with all their derivatives up to the boundary $\partial\Omega_\pm$ (including $\infty$), invertible everywhere in the closed regions, and such that for $\lambda\in\Sigma$,

$$g(\lambda)=g_+(\lambda)\,g_-(\lambda) \tag{3.17}$$

and

$$\lim_{\substack{\lambda\to\infty \\ \lambda\in\Omega_{m+1}}} g_+(\lambda) = \lim_{\substack{\lambda\to\infty \\ \lambda\in\Omega_{-m}}} g_-(\lambda) \tag{3.18}$$

As in Section 2, such a factorization is unique, if it exists. We let $\widetilde{G}$ be the set of all $g \in G$ for which such a factorization exists, and as before $\widetilde{G}$ with group law

$$g * h = g_+ \, h \, g_- \tag{3.19}$$

has Lie-algebra $\underset{\sim}{\tilde{g}}$. The Hamiltonian coadjoint orbit theory may now be developed in analogy with Section 2, but we leave the details to the reader and focus on the relationship between the flows and the factorization (3.17), (3.18).

Let $f$ be a function from $\mathbb{C} \to \mathbb{C}$ and suppose that

$$e^{tf(A^0(\lambda))} = g_+(t,\lambda)\, g_-(t,\lambda) \tag{3.20}$$

is a factorization of the above type for some $A^0 \in \underset{\sim}{g}^*$, $f(A^0(\cdot)) \in \underset{\sim}{g}$. Differentiation with respect to $t$ shows that for

$$A(t,\lambda) \equiv g_+^{-1}(\lambda,t)\, A^0(\lambda)\, g_+(\lambda,t) \, , \tag{3.21}$$

$$f(A(t,\lambda)\,) = g_+^{-1}\dot{g}_+ + \dot{g}_- g_-^{-1} \, . \tag{3.22}$$

A simple computation using (iii) shows that $g_+^{-1}\dot{g}_+$ and $\dot{g}_- g_-^{-1}$ satisfy (iii)$'$ and hence lie in $\underset{\sim}{q}$. By Cauchy we find

$$(\pi_- g_+^{-1}\dot{g}_+)(\lambda) = \frac{1}{2} g_+^{-1}\dot{g}_+(x_m + i\infty) \tag{3.23}$$

and

$$(\pi_- \dot{g}_- g_-^{-1})(\lambda) = \dot{g}_- g_-^{-1}(\lambda) - \frac{1}{2}\dot{g}_- g_-^{-1}(x_{-m} + i\infty) \tag{3.24}$$

But from (3.18),

$$g_+^{-1}\dot{g}_+(x_m + i\infty) = \dot{g}_- g_-^{-1}(x_{-m} + i\infty) \, ,$$

and so

$$\pi_- f(A(t,\cdot)) = \dot{g}_- g_-^{-1}(\lambda) \, , \tag{3.25}$$

and we find by the familiar calculation that $A(t,\lambda)$ solves

$$\begin{aligned}\frac{d}{dt}A(t,\lambda) &= \left[(\pi_- f(A(t,\cdot)))(\lambda)\,,\, A(t,\lambda)\right]\,,\\ A(0,\lambda) &= A^0(\lambda)\,.\end{aligned} \tag{3.26}$$

(Of course (3.26) follows abstractly from (3.20) and (3.18) by $R$-matrix theory, but we have included the above direct computation for the convenience of the reader who wants to see "how it works".) In what follows we will be interested in particular in the case $e^{t\log(A^0(\lambda))}$, where $A^0 = \widetilde{M}_0(\lambda)$.

Given $M_0$, we consider the general case (1.19) subject as above to (3.1). Every such decomposition $S = S_+ \cup S_-$ gives rise to a skeleton in a natural way. For example if $S$ has the shape

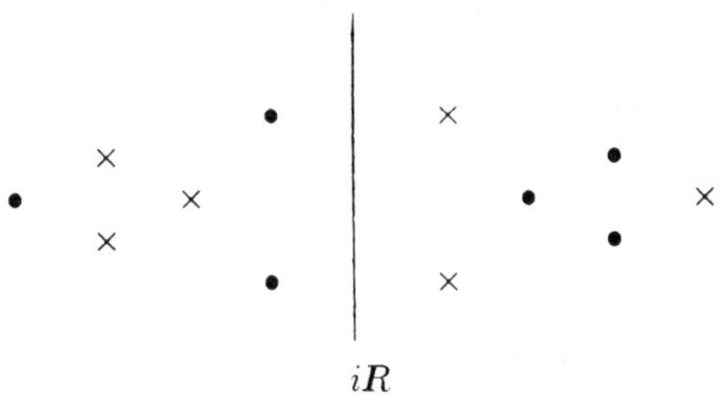


Figure 3.4

($\times$ lies in $S_+$, $\bullet$ lies in $S_-$), then we obtain the skeleton $\Sigma = \bigcup_{j=-2}^{2} \Sigma_j$

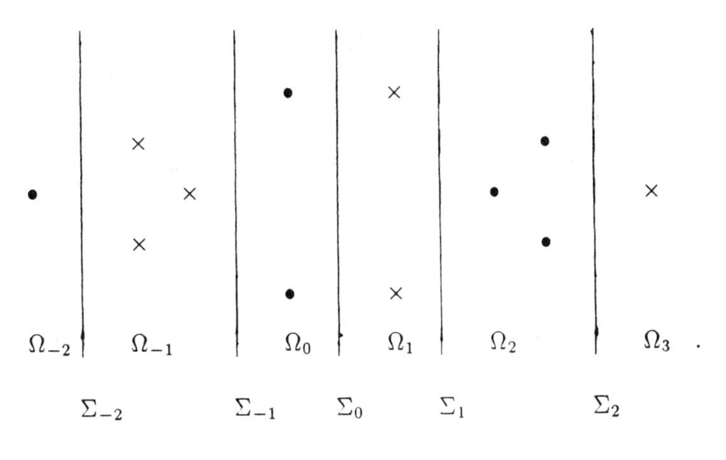


Figure 3.5

Moreover, if we attach signs to $\Omega_j$ in the natural way, $\Omega_+ = \Omega_{-1} \cup \Omega_1 \cup \Omega_3$, $\Omega_- = \Omega_{-2} \cup \Omega_0 \cup \Omega_2$, then the decomposition $\mathbb{C} \setminus \Sigma = \Omega_+ \cup \Omega_-$ agrees precisely with (3.5). On the other hand, if we interchange $\times \leftrightarrow \bullet$, then the signs of the $\Omega_j$'s reverse. Thus every $S = S_+ \cup S_-$ determines a skeleton $\Sigma$ and a decomposition $\mathbb{C} = \Omega_+ \cup \Omega_-$ in which the signs either agree with (3.5), or are the exact opposite. We will restrict ourselves in this section to the case where the signs agree with (3.5): the other case is similar and presents no new difficulties (note also that a factorization $g^{-1} = g_+ g_-$ gives rise to a factorization $g = g_-^{-1} g_+^{-1}$ with the opposite sign).

By inspection we see that the Moser-Veselov factorization

$$A^0(\lambda) = \frac{I - \lambda M_0 - \lambda^2 J^2}{1 - \lambda^2} = \left(\frac{\omega^T + \lambda J}{1 + \lambda}\right)\left(\frac{\omega - \lambda J}{1 - \lambda}\right) \tag{3.27}$$

corresponds exactly to the factorization

$$A^0(\lambda) = e^{t \log A^0(\lambda)}\Big|_{t=1} = g_+(1, \lambda)\, g_-(1, \lambda)\ . \tag{3.28}$$

(Technical remark: we assume without loss of generality that $1 - \lambda \neq 0$ on $\overline{\Omega}_-$; otherwise replace $1 - \lambda$ with $1 - \epsilon\lambda$ for suitable $\epsilon > 0$, etc.) From (3.21) and (3.28),

$$\begin{aligned} A(1, \lambda) &= g_+(1, \lambda)^{-1}\, A^0(\lambda)\, g_+(1, \lambda) \\ &= g_-(1, \lambda)\, g_+(1, \lambda)\ , \end{aligned}$$

which shows that the flow generated by $e^{t \log A^0(\lambda)}$ interpolates the discrete Euler-Arnold equations at integer times. As in Section 2 the flow is Hamiltonian on $\underline{g}^*$, and one finds once again that

$$A(t, \lambda) = \frac{I - \lambda M(t) - \lambda^2 J^2}{1 - \lambda^2}\ . \tag{3.29}$$

The branches of the logarithm in $g(t, \lambda) = e^{t \log A^0(\lambda)}$ are chosen such that $g(t, -\bar{\lambda})^* = g(t, \lambda)$ (see (3.35) below). This implies

$$g_+(t, \bar{\lambda})^* = g_-(t, \lambda) \tag{3.30}$$

by uniqueness, and in turn,

$$M(t) + M^*(t) = 0 \tag{3.31}$$

Thus the interpolating flow generated by $e^{t \log A^0(\lambda)}$ induces a flow $t \to M(t)$ on the skew-adjoint matrices. We leave it to the reader to show as before that the flow is Hamiltonian and integrable.

In general we do not have $\overline{g(t, \bar{\lambda})} = g(t, \lambda)$ and so $M(t)$ is not real for all $t$, even though $M_0 = M(0)$ is real. However for $t = k \in \mathbb{Z}$, $g(k, \lambda) = e^{k \log A^0(\lambda)} = (A^0(\lambda))^k$, so $\overline{g(k, \bar{\lambda})} = g(k, \lambda)$ and hence $M(k)$ is real. Thus the flow intersects the real part of the manifold of skew-adjoint matrices at integer times (as it should). A similar situation occurs in the interpolating flow for the classical Cholesky algorithm if the matrix $M(0)$ has (an odd number of) negative eigenvalues in which case $\log M(0)$ cannot be chosen real.

It remains to show that the desired factorization

$$g(\lambda) = g(t, \lambda) = e^{t \log A^0(\lambda)} = g_+(t, \lambda)\, g_-(t, \lambda) \tag{3.32}$$

exists. Note first that $A^0(\lambda)$ is smooth and det $A^0(\lambda) \neq 0$ on $\Sigma$, by construction: hence $\log A^0(x_j + iy)$ can be determined on each $\Sigma_j$ by continuity and we can, and do, choose $\log A^0(x_j + i\infty)$ to be real for each $j$. As $A^0(x_j + i\infty) = A^0(x_j - i\infty)$, we must have

$$\log A^0(x_j + i\infty) - \log A^0(x_j - i\infty) = 2\pi i \operatorname{diag}(k_1, \ldots, k_N)$$

for suitable integers $k_i = k_i(j)$. Now for $\lambda \in \Sigma$, $B(\lambda) \equiv (\log A^0)^*(-\bar{\lambda})$ is a logarithm of $A^0(\lambda)$: indeed $e^{B(\lambda)} = e^{(\log A^0)^*(-\bar{\lambda})} = A^0(-\bar{\lambda})^* = A^0(\lambda)$, and as $B(x_j + i\infty)$ is also real, we must have

$$\log A^0(\lambda) = (\log A^0)^*(-\bar{\lambda}) \ . \tag{3.33}$$

This implies

$$\begin{aligned}
&\log A^0(x_j+i\infty)-\log A^0(x_j-i\infty)+\log A^0(x_{-j}+i\infty)-\log A^0(x_{-j}-i\infty) \\
&=\log A^0(x_j+i\infty)-\log A^0(x_j-i\infty)+(\log A^0)^*(x_j+i\infty)-(\log A^0)^*(x_j-i\infty) \\
&=2\pi i\ \mathrm{diag}(k_1,\ldots,k_N)+\overline{2\pi i\ \mathrm{diag}(k_1,\ldots,k_N)} \\
&=0\,.
\end{aligned} \tag{3.34}$$

Using (3.33) it is now easy to show that $\log A^0(\lambda)$ satisfies the zero winding condition (iii)$'$, and hence $e^{t\log A^0(\lambda)}\in G$. The above calculations also show that

$$(e^{t\log A^0(\cdot)})^*(-\bar{\lambda})=(e^{t\log A^0(\cdot)})(\lambda)\,. \tag{3.35}$$

To apply standard factorization techniques (see, e.g. [G-K]), we must reduce the problem to the case where $g(\lambda)\to I$ as $\lambda\to\infty$. We do this as follows.

For $-m\le k\le m+1$, set

$$S_k^{\pm}=g(x_{k-1}\pm i\infty)\,g(x_{k-2}\pm i\infty)^{-1}\,g(x_{k-3}\pm i\infty)\,g(x_{k-4}\pm i\infty)^{-1}\cdots I \tag{3.36}$$

($S_{-m}^{\pm}\equiv I$). By (iii),

$$S_{m+1}^{+}=S_{m+1}^{-}\,. \tag{3.37}$$

For each $-m\le k\le m+1$, define a piecewise analytic function

$$\hat{g}(\lambda)=S_k^{+}\,e^{\gamma(\lambda)(\log S_k^{-}-\log S_k^{+})}\,,\qquad \lambda\in\Omega_k\,, \tag{3.38}$$

where $\gamma$ has properties (i)$''$, (ii)$''$ above and any branch of the logarithm will do, provided $\log S_{-m}^{+}=\log S_{-m}^{-}$ and $\log S_{m+1}^{+}=\log S_{m+1}^{-}$, so that

$$\hat{g}\mid\Omega_{-m}=I\qquad\text{and}\qquad \hat{g}\mid\Omega_{m+1}=S_{m+1}^{+}=S_{m+1}^{-}\,. \tag{3.39}$$

We have

$$\hat{g}(\lambda)\to S_k^{\pm} \tag{3.40}$$

as Im $\lambda \to \pm\infty$ respectively in $\overline{\Omega}_k$. Finally, $\hat{g}(\lambda)$ is clearly invertible in $\overline{\Omega}_k$.

Define

$$\hat{g}_\pm \equiv \hat{g} \mid \Omega_\pm \ , \tag{3.41}$$

and we denote the continuations of $\hat{g}_\pm$ to the boundary by the same symbols.

Now the point is this: as the reader may check, $S_k^\pm$ have been chosen precisely in such a way that

$$g(x, \pm i\infty) = \hat{g}_+(x_j \pm i\infty)\,\hat{g}_-(x_j \pm i\infty) \ . \tag{3.42}$$

For $\lambda \in \Sigma$, set

$$g^\#(\lambda) = (\hat{g}_+(\lambda))^{-1}\, g(\lambda)(\hat{g}_-(\lambda))^{-1} \ . \tag{3.43}$$

By (3.42),

$$g^\#(\lambda) \to I \tag{3.44}$$

as $\lambda \to \infty$ on $\Sigma$. Furthermore, if we can produce a factorization

$$g^\# = g_+^\#(\lambda)\, g_-^\#(\lambda) \tag{3.45}$$

with $g_\pm^\#(\lambda) \to I$ as $\lambda \to \infty$, then

$$g_+(\lambda) = \hat{g}_+(\lambda)\, g_1^\#(\lambda)\, D \ , \tag{3.46}$$

$$g_-(\lambda) = D^{-1}\, g_-^\#(\lambda)\, \hat{g}_-(\lambda) \ , \tag{3.47}$$

provide the desired factorization $g = g_+ g_-$, for an appropriately chosen constant, invertible diagonal matrix $D$. Indeed, we must choose $D$ such that

$$\hat{g}_+(x_m + i\infty)D = D^{-1}\, \hat{g}_-(x_{-m} + i\infty) \ ,$$

i.e.,

$$D = \sqrt{(S_{m+1}^+)^{-1}} = \sqrt{(S_{m+1}^-)^{-1}} > 0 \tag{3.48}$$

by (3.39). (Note that $S_{m+1}^+ = S_{m+1}^- > 0$ as $g \in G$).

The factorization problem (3.45) can be solved by standard techniques, and we sketch the approach for the convenience of the reader (see [G-K]). Write

$$\begin{aligned} g_+^{\#} &= I + h_+ \\ (g_-^{\#})^{-1} &= I + h_- \end{aligned} \tag{3.49}$$

and (3.45) takes the form

$$g^{\#} h_- = (h_+ + (I - g^{\#}))\ , \tag{3.50}$$

where $h_-$ and $h_+ + (I - g^{\#})$ belong to $L^2(\Sigma)$. Now the operators $\pi_\pm$ above, restricted to $L^2(\Sigma)$, become complementary proejctions, $\pi_+ + \pi_- = I$, $\pi_\pm^2 = \pi_\pm$, $\pi_+\pi_- = 0$, and (3.50) can be rewritten as

$$\begin{pmatrix} \pi_+ g^{\#} \pi_+ & \pi_+ g^{\#} \pi_- \\ \pi_- g^{\#} \pi_+ & \pi_- g^{\#} \pi_- \end{pmatrix} \begin{pmatrix} \mu_+ \\ \mu_- \end{pmatrix} = \begin{pmatrix} \nu_+ \\ \nu_- \end{pmatrix} \tag{3.51}$$

for suitable $\mu_\pm, \nu_\pm \in \pi_\pm(L^2(\Sigma))$. As $g^{\#}$ is invertible on $\Sigma$, the matrix operator in (3.51) is a bijection: on the other hand, $\pi_\pm g^{\#} \pi_\pm = \pi_\pm (g^{\#} - I)\pi_\mp$ are compact on $L^2(\Sigma)$ by standard computations. Hence the diagonal operator

$$\begin{pmatrix} \pi_+ g^{\#} \pi_+ & 0 \\ 0 & \pi_- g^{\#} \pi_- \end{pmatrix}$$

is Fredholm of index zero. Thus, if we can show that

$$\ker\, \pi_+ g^{\#} \pi_+ = 0 \quad \text{and} \quad \ker\, \pi_- g^{\#} \pi_- = 0\ , \tag{3.52}$$

then, in particular, the problem

$$\pi_- g^{\#} \pi_- \mu_- = \nu_- \tag{3.53}$$

has a unique solution $\mu_-$ in $\pi_- L^2(\Sigma)$ for any $\nu_- \in \pi_- L^2(\Sigma)$. Setting

$$\nu_- = \pi_-(I - g^{\#})\ , \tag{3.54}$$

we obtain from (3.53)

$$g^{\#} \mu_- = I - g^{\#} + \mu_+$$

for some $\mu_+ \in \pi_+ L^2(\Sigma)$, i.e.

$$g^\#(I + \mu_-) = (I + \mu_+)$$

and

$$g_+^\# = I + \mu_+ , \qquad g_-^\# = (I + \mu_-)^{-1} \tag{3.55}$$

are the desired factors (for the invertibility of $(I + \mu_\pm)$, see below).

It remains to prove the kernel conditions (3.52). Note that

$$\pi_- g^\# \pi_- \mu_-^\# = 0 \Rightarrow \mu_-^\# = 0 \quad \text{for} \quad \mu_-^\# \in \pi_- L^2(\Sigma)$$

$$\Leftrightarrow$$

$$g^\# \mu_-^\# = \mu_+^\# \Rightarrow \mu_+^\#, \mu_-^\# = 0 \quad \text{for} \quad \mu_\pm^\# \in \pi_\pm L^2(\Sigma) \tag{3.56}$$

$$\Leftrightarrow$$

$$g\mu_- = \mu_+ \Rightarrow \mu_+, \mu_- = 0 \quad \text{for} \quad \mu_\pm \in \pi_\pm L^2(\Sigma)$$

(see (3.43)).

Similarly

$$\ker \pi_+ g^\# \pi_+ = 0$$

$$\Leftrightarrow \tag{3.57}$$

$$g^{-1}\mu_- = \mu_+ \Rightarrow \mu_+, \mu_- = 0 \quad \text{for} \quad \mu_\pm \in \pi_\pm L^2(\Sigma) .$$

So suppose

$$g\mu_- = \mu_+ , \qquad \mu_\pm \in \pi_\pm L^2(\Sigma) .$$

By Cauchy,

$$\begin{aligned}
0 &= \int_{\Sigma_m} \mu_-^*(-\bar\lambda)\, \mu_+(\lambda)\, \bar{d}\lambda \\
&= \int_{\Sigma_m} \mu_-(-\bar\lambda)\, g(\lambda)\, \mu_-(\lambda)\, \bar{d}\lambda \\
&= \int_{\Sigma_m} \mu_-^*(-\bar\lambda)\, g^*(-\bar\lambda)\, \mu_-(\lambda)\, \bar{d}\lambda , \quad \text{by (3.35)}, \\
&= \int_{\Sigma_m} \mu_+^*(-\bar\lambda)\, \mu_-(\lambda)\, \bar{d}\lambda \\
&= \int_{\Sigma_{m-1}} \mu_+^*(-\bar\lambda)\, \mu_-(\lambda)\, \bar{d}\lambda , \quad \text{by Cauchy}, \\
&= \cdots = \int_{\Sigma_0} \mu_\mp^*(-\bar\lambda)\, \mu_\pm(\lambda)\, \bar{d}\lambda ,
\end{aligned}$$

where the signs depend on whether $m$ is even or odd. Thus, as $\lambda = -\bar{\lambda}$ on $\Sigma_0$, we obtain either

$$0 = \int_{\Sigma_0} \mu_-^*(\lambda)\, g(\lambda)\, \mu_-(\lambda)\, \bar{d}\lambda$$

or

$$0 = \int_{\Sigma_0} \mu_-^*(\lambda)\, g^*(\lambda)\, \mu_-(\lambda)\, \bar{d}\lambda\ .$$

But for $\lambda \in \Sigma_0$,

$$g(\lambda) = g(\lambda)^* = e^{t \log A^0(\lambda)}$$

is positive definite, and hence $\mu_-(\lambda) = 0$ on $\Sigma_0$, which implies $\mu_-(\lambda) = \mu_+(\lambda) = 0$ everywhere. The same argument works for $g^{-1}$ as $g^{-1}(-\bar{\lambda})^* = g^{-1}(\lambda)$. This completes the proof of the kernel conditions (3.52).

**Remark.** The above proof of (3.52) is patterned after the so called Vanishing Lemmas in [DT], [DTT] and in [BDT].

To show that $(I + \mu_\pm)$ in (3.55) are invertible, suppose that $\det(I + \mu_+)$, say, vanishes at a point $\lambda^0 \in \overline{\Omega}_+$. Then

$$f_\pm(\lambda) = \frac{\det\, (I + \mu_\pm(\lambda))}{\lambda - \lambda^0} (\det\, \hat{g}_\pm(\lambda)\,)^{\pm 1}$$

gives a solution of the scalar factorization problem

$$(\det\, g)(\lambda)\, f_-(\lambda) = f_+(\lambda)\ , \quad \lambda \in \Sigma\ ,$$

with $f_\pm \in \pi_\pm L^2(\Sigma)$. As $\overline{(\det\, g)(-\bar{\lambda})} = \det\, g(\lambda)$ and as $(\det\, g)(\lambda) > 0$ for $\lambda \in \Sigma_0$, the above "Vanishing Lemma" argument now applies and we conclude that $\det(I + \mu_\pm(\lambda)\,) \equiv 0$, which is a contradiction. Hence $I + \mu_\pm$ are invertible.

This completes the proof of our results, which we summarize in the following theorem:

**Theorem 3.58.** *Let* $A^0(\lambda) = (I - \lambda M_0 - \lambda^2 J^2)/(1 - \lambda^2)$, *and let* $g_+(t, \lambda), g_-(t, \lambda)$ *solve the factorization problem*

$$e^{t \log A^0(\lambda)} = g_+(t, \lambda)\, g_-(t, \lambda)\ ,$$

$$\lim_{\lambda \to \infty,\, \lambda \in \Omega} g_+(t, \lambda) = \lim_{\lambda \to 0,\, \lambda \in \Omega_{-m}} g_-(t, \lambda)\ ,$$

*for a given decomposition* (1.19).

*Then*

$$A(t,\lambda) = g_+(t,\lambda)^{-1} A_0(\lambda)\, g_+(t,\lambda)$$

*interpolates the Moser-Veselov algorithm (for the given decomposition* (1.19)) *at integer times, and solves the (Hamiltonian) equation*

$$\frac{d}{dt}A(t,\lambda) = \left[(\pi_- \log A(t,\cdot))(\lambda)\ ,\ A(t,\lambda)\right] .$$

*For all* $t$, $A(t,\lambda)$ *has the form*

$$A(t,\lambda) = (I - \lambda M(t) - \lambda^2 J^2)\,/\,(1-\lambda^2)$$

*where* $M(t)$ *is skew-adjoint, and the flow* $t \longrightarrow M(t)$ *induced on the skew-adjoint matrices, is Hamiltonian and integrable.* □

**Remark 3.59.** Suppose that (3.1) fails. Then we may have, for example, a decomposition $S = S_+ \cup S_-$ of the form,

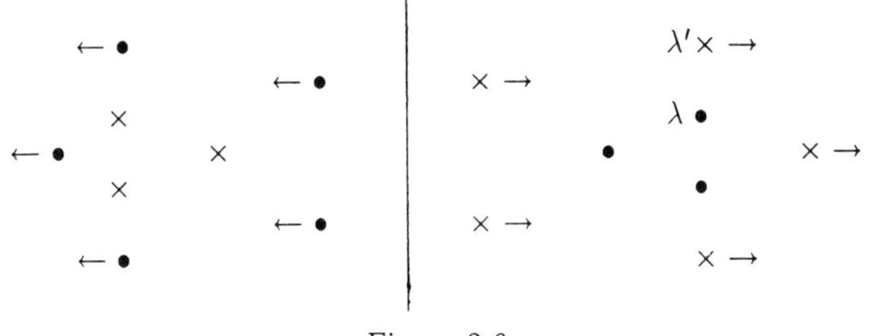


Figure 3.6

for which Re $\lambda =$ Re $\lambda'$, but $\lambda \in S_-$ and $\lambda' \in S_+$. A convenient way to accommodate this situation is to imagine the elements in $S_+ \cap \{\text{Re }\lambda > 0\}$ shifted infinitesimally to the right, and the elements in $S_- \cap \{\text{Re }\lambda < 0\}$ shifted correspondingly to the left, as indicated in the figure. This leads to a decomposition $S' = S'_+ \cup S'_-$ satisfying (3.1), which leads in turn to a skeleton $\Sigma'$ consisting of lines parallel to $iR$ as in (3.2) above. If we now let the displaced elements return to their original positions, constantly deforming $\Sigma'$ so as to avoid contact with $S^1$, then we are led to a skeleton $\Sigma = \bigcup_{j=-2}^{2} \Sigma_j$ of the form

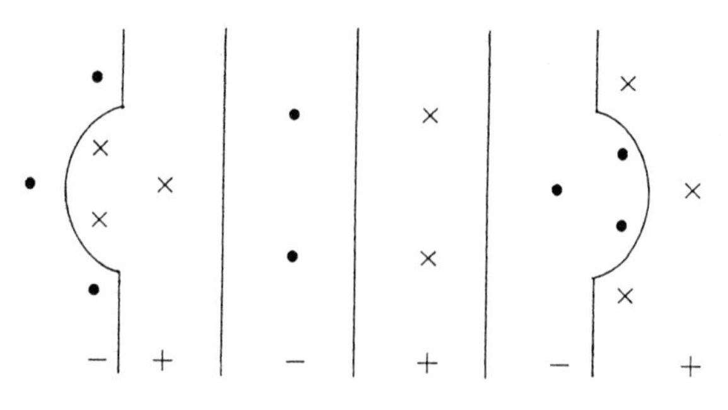


Figure 3.7

As indicated in the figure, signatures are attached to the components of $\Omega = \mathbb{C} \setminus \Sigma$ as before, and $\Sigma$ is shown here with the $(+)$-orientation. Note also that $\Sigma$ can, and should, be chosen such that

$$\Sigma = \overline{\Sigma} \quad \text{and} \quad \Sigma = -\Sigma \ . \tag{3.60}$$

We leave it to the reader to verify that all the results and computations of this section extend verbatim to deformed skeletons $\Sigma$ of the above type. This completes the analysis of the discrete Euler-Arnold equation in the general case.

## 4. Billiards in an elliptical region.

### (a) The factorization problem and the interpolating flow.

In [MV] the authors consider the billiard map in an elliptical region

$$E = \{x : (x, C^{-2}x) \leq 1\}$$

in $R^N$, where $C$ is positive and diagonal. If a billiard ball strikes $\partial E$ at a point $x_0$ from a direction $y_0$, $\|y_0\| = 1$, then after normal reflection ("angle of incidence = angle of reflection") the ball strikes $\partial E$ at a second point $x_1$ from a direction $y_1$,

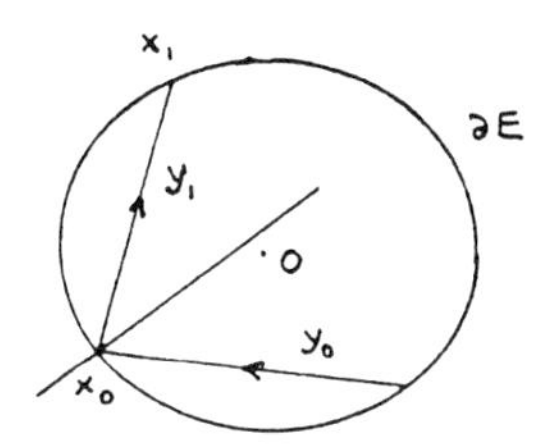


Figure 4.1

The billiard map is the map $\Psi_b$ from $Y = \partial E \times \{\|y\| = 1\}$ to $Y$ taking $(x_0, y_0) \mapsto (x_1, y_1)$. It is a remarkable discovery in [MV] (see also [V]) that $\Psi_b$ can be solved (up to a sign — see below) in terms of a factorization problem as in Section 2. Consider the matrix polynomial

$$L_0(\lambda) = y_0 \otimes y_0 + \lambda x_0 \wedge y_0 - \lambda^2 C^2 , \quad (x_0, y_0) \in Y , \tag{4.1}$$

introduced in connection with geodesic flow on an ellipsoid in [M]. The matrix $L_0$ can be factored

$$L_0(\lambda) = (\lambda C + y_0 \otimes \xi_0)(-\lambda C + \xi_0 \otimes y_0) \tag{4.2}$$

with $\xi_0 = C^{-1}x_0$, $\|\xi_0\| = 1$. Exchanging factors, we obtain

$$L_0'(\lambda) = (-\lambda C + \xi_0 \otimes y_0)(\lambda C + y_0 \otimes \xi_0) \tag{4.3}$$

which can again be factored (uniquely) in the form

$$L_0'(\lambda) = (\lambda C + y_0' \otimes \xi_0')(-\lambda C + \xi_0' \otimes y_0') \tag{4.4}$$

with $y_0' = \xi_0$, $\|y_0'\| = 1$ and $\|\xi_0'\| = 1$. Then the map $(x_0, y_0) \to (C\xi_0', y_0')$ agrees with the "skew-hodograph" mapping $\phi$ (our $\phi$ differs from the skew-hodograph mapping in [MV] by an overall minus sign),

$$\begin{aligned} \phi : Y &\to Y \\ (x_0, y_0) &\mapsto (-Cy_1, C^{-1}x_0)\ , \end{aligned} \tag{4.5}$$

and the square of $\phi$ gives $\Psi_b$ up to a sign, i.e.

$$\phi^2 = \phi \circ \phi = -\Psi_b \tag{4.6}$$

and

$$\phi^4 = \Psi_b^2 \tag{4.7}$$

(see [MV]).

From the above formulae, it is clear how to express the billiard ball problem in terms of the loop group of Section 2. We consder the coadjoint orbit through

$$A^0(\lambda) = L_0(\lambda)/(1-\lambda^2) = \frac{y_0 \otimes y_0 + \lambda x_0 \wedge y_0 - \lambda^2 C^2}{1-\lambda^2}\ , \tag{4.8}$$

and consider the (Hamiltonian) flow $A(t,\lambda)$ generated by a factorization problem

$$\begin{aligned} e^{t \log A^0(\lambda)} &= g_+(t,\lambda)\, g_-(t,\lambda) \\ g_+(t,\infty) &= g_-(t,\infty) \end{aligned}\ , \tag{4.9}$$

where the analytic properties of $g_\pm(t,\cdot)$ reflect the half-plane splitting $S_+ \cup S_-$ of the roots of $\det(y_0 \otimes y_0 + \lambda x_0 \wedge y_0 - \lambda^2 C^2)$. As before we find

$$\begin{aligned} A(t=1,\lambda) &= g_+(t,\lambda)^{-1} A^0(\lambda)\, g_+(t,\lambda)\big|_{t=1} \\ &= L_0'(\lambda)/(1-\lambda^2)\ , \end{aligned} \tag{4.10}$$

and so $A(t=2,\lambda)$ implements $\Psi_b$ up to a sign.

There are significant obstacles, however, in trying to place the above scheme on a rigorous basis. The source of the difficulty is that $\det(L_0(\lambda))$ vanishes at $\lambda = 0$. This implies, first, that a splitting $S = S_+ \cup S_-$ of the form (1.19) does not in fact exist. Second, $e^{t \log A^0(\lambda)}$ either blows up or fails to be invertible at $\lambda = 0$ (for $t > 0$, the latter occurs — see below).

To overcome these difficulties note first that by the Weinstein-Aronszajn formulae (see e.g. [K], p. 244),

$$\det L_0(\lambda) = (-1)^N (\det C^2)\, \lambda^{2N-2}(\lambda^2 - (C^{-2}x_0, y_0)^2)\,. \tag{4.11}$$

In particular, as $L_0(iy) > 0$ for all $|y|$ sufficiently large, $L_0(\lambda) \geq 0$ for all $\lambda \in iR$, and hence for any $\delta > 0$ $\det(L_0(\lambda)+\delta) > 0$ on $iR$, and as $\det(L_0(\lambda)+\delta) = \det(L_0(-\lambda)+\delta)$ it follows that $S^\delta = \{\lambda : \det(L_0(\lambda)+\delta) = 0\}$ has decompositions $S^\delta = S_+^\delta \cup S_-^\delta$ satisfying (1.19). One such decomposition that exists for all $\delta > 0$ is $S_+^\delta = S^\delta \cap \{\mathrm{Re}\,\lambda > 0\}$, $S_- = S^\delta \cap \{\mathrm{Re}\,\lambda < 0\}$, which gives rise (as explained in Section 3) to the skeleton

$$\Sigma = iR \tag{4.12}$$

and the decomposition

$$\mathbb{C} \setminus \Sigma = \Omega_+ \cup \Omega_- = \{\mathrm{Re}\,\lambda > 0\} \cup \{\mathrm{Re}\,\lambda < 0\}\,. \tag{4.13}$$

In what follows we will assume that $\Sigma$ and $\Omega_\pm$ are given by (4.12) and (4.13) respectively, as in Section 2. As noted above $A^0(\lambda) \geq 0$ on $\Sigma$; for $t > 0$ this implies that $e^{t \log A^0(\lambda=0)}$ is not invertible, which makes the factorization (4.9) singular at $\lambda = 0 \in \Sigma$.

To proceed, we need the following lemma. Recall from regular perturbation theory that on $\Sigma$ the eigenvalues and the (orthonormalized) eigenvectors of the (real anlaytic, self-adjoint) matrix $A^0(\lambda)$, are real-analytic. Also, $C^{-2}x$ is the outward normal to $\partial E$ at $x$, and so, by the geometry of Fig. 4.1, the mapping $\Psi_b$ acts naturally on $Y_+ = Y \cap \{(x,y) : (x, C^{-1}y) > 0\}$.

**Lemma 4.14.** *The eigenvalues of $A^0(\lambda)$, $(x_0, y_0) \in Y_+$, do not vanish on $\Sigma \setminus 0$. If an eigenvalue $\mu(\lambda)$ vanishes at $\lambda = 0$, then it vanishes precisely to second order,*

$$\mu(\lambda) = \lambda^2 c_0(1 + o(1)) , \quad \textit{as} \quad \lambda \to 0 , \tag{4.14}$$

*where $c_0 < 0$.*

**Proof:** We must verify (4.14). It is enough to consider the eigenvalues $\tilde{\mu} = (1 - \lambda^2)\mu$ of $L_0$,

$$(y_0 \otimes y_0 + i\gamma x_0 \wedge y_0 + \gamma^2 C^2)\, u(\gamma) = \tilde{\mu}(\gamma)\, u(\gamma) , \quad \gamma \in \boldsymbol{R} . \tag{4.15}$$

As $\tilde{\mu}(0) = 0$, we must have $(y_0, u(0)) = 0$ as $L_0(0)u(0) = 0$. From the formulae $\tilde{\mu}(\gamma) = (\overline{u(\gamma)}, L_0(i\gamma)\, u(\gamma))$ and $(\bar{u}(\gamma), u(\gamma)) = 1$, we obtain

$$\dot{\tilde{\mu}}(\gamma) = (\overline{u(\gamma)}, (ix_0 \wedge y_0 + 2\gamma C^2)\, u(\gamma)) .$$

In particular,

$$\dot{\tilde{\mu}}(0) = 0$$

and

$$\ddot{\tilde{\mu}}(0) = 2((C\overline{u(0)}, Cu(0)) - \operatorname{Re}(i(\overline{\dot{u}(0)}, y)(x, u(0)))) .$$

But from $(L_0(i\gamma) - \tilde{\mu}(\gamma))\, \dot{u}(\gamma) + (ix_0 \wedge y_0 + 2\gamma C^2 - \dot{\tilde{\mu}}(\gamma))\, u(\gamma) = 0$, we find

$$y_0 \otimes y_0\, \dot{u}(0) + i\, x_0 \wedge y_0\, u(0) = 0 ,$$

which implies

$$(y_0, \dot{u}(0)) = i(x_0, u(0)) ,$$

and hence

$$\ddot{\tilde{\mu}}(0) = -c_0 ,$$

where

$$c_0 = 2(|(x, u(0))|^2 - (C\overline{u(0)}, Cu(0))) . \tag{4.16}$$

But $|(x_0, u(0))| = |(C^{-1}x_0, C\,u(0))| \leq \|C^{-1}x_0\|(C\overline{u(0)}, Cu(0))^{1/2} = (C\overline{u(0)}, Cu(0))^{1/2}$, and so $c_0 \leq 0$, with equality if and only if $C^{-1}x_0 = \beta(u(0))$ for some scalar $\beta$, which implies in turn

$$(y_0, C^{-1}x_0) = \beta(y_0, u(0)) = 0$$

contradicting $(x_0, y_0) \in Y_+$. Thus $c_0 < 0$ and (4.14) is proved. □

**Remark 4.17.** The quantity $(x, C^{-2}y)$ is conserved under $\Psi_b$. In fact a simple calculation shows that $(x, C^{-2}y)$ is preserved under the skew-hodograph mapping $\phi$.

Our first goal is to prove the following result. Let $(x, y) \in Y_0$ and set $A(\lambda) = (y \otimes y + \lambda\, x \wedge y - \lambda^2 C^2)/(1 - \lambda^2)$. Also let $t > 0$.

**Theorem 4.18.** *There exists a factorization*

$$e^{t \log A(\lambda)} = g_+(t, \lambda)\, g_-(t, \lambda)\,, \quad \lambda \in iR \setminus 0 \tag{4.19}$$

*with the properties*

(i) $g_-(t, \lambda) = g_+^*(t, \lambda)$, $\lambda \in iR \setminus 0$;

(ii) $g_\pm(t, \lambda)$ have analytic extensions to $\operatorname{Re} \lambda > 0$, $\operatorname{Re} \lambda < 0$ *respectively*;

(iii) $g_\pm(t, \lambda)$ are continuous and bounded in $\{\operatorname{Re} \lambda \geq 0\} \setminus 0$, $\{\operatorname{Re} \lambda \leq 0\} \setminus 0$ *respectively;*

(iv) *as* $\lambda \to 0$ *in* $\operatorname{Re} \lambda > 0$, $\operatorname{Re} < 0$ *respectively.*

$$0 < \eta \leq \left| \frac{\det g_\pm(t, \lambda)}{\lambda^{(N-1)t}} \right| \leq \eta^{-1} < \infty \tag{4.20}$$

*for some constant* $\eta$ *independent of* $t$,

*and*

(v) $g_\pm(t, \lambda) \to C^t$ *as* $\lambda \to \infty$.

*Moreover this factorization is unique.*

**Remark.** Condition (iv) implies in particular that $g_\pm(t, \lambda)$ are invertible in $\{\operatorname{Re} \lambda \geq 0\} \setminus 0$, $\{\operatorname{Re} \lambda \leq 0\} \setminus 0$ respectively.

For $t = 1$, the factorziation in (4.19) agrees exactly with the Moser-Veselov factorization,

$$A^0(\lambda) = \left( \frac{\lambda C + y_0 \otimes \xi_0}{\lambda + 1} \right) \left( \frac{-\lambda C + \xi_0 \otimes y_0}{-\lambda + 1} \right), \tag{4.21}$$

so that

$$g_+(1,\lambda) = \left(\frac{\lambda C + y_0 \otimes \xi_0}{\lambda + 1}\right), \quad g_-(1,\lambda) = \left(\frac{-\lambda C + \xi_0 \otimes y_0}{-\lambda + 1}\right) \tag{4.22}$$

and hence, by the familiar computation,

$$A(\lambda, t) = g_+(t,\lambda)^{-1} A^0(\lambda) \, g_+(t,\lambda)$$

interpolates the skew-hodograph transformation at integer times. To verify (4.22) note that properties (i), (ii), (iii) and (v) above are immediate for the factors in (4.21). Again by the Weinstein-Aronszajn formulae,

$$\det\left(\frac{\lambda C + y_0 \otimes \xi_0}{\lambda + 1}\right) = \frac{\lambda^{N-1}}{(\lambda+1)^N}(\det C)(\lambda + (x_0, C^{-2} y_0)) \tag{4.23$_+$}$$

and

$$\det\left(\frac{-\lambda C + \xi_0 \otimes y_0}{-\lambda + 1}\right) = \frac{(-\lambda)^{N-1}}{(-\lambda+1)^N}(\det C)(-\lambda + (x_0, C^{-2} y_0))\ , \tag{4.23$_-$}$$

which establishes (iv), and (4.22) now follows by uniqueness.

To prove Theorem 4.18 we begin with the strictly positive matrix $A(\lambda) + \delta$, $\lambda \in iR$ and $\delta > 0$. By the results of Section 2, there exists a (unique) factorization

$$e^{t \log(A(\lambda)+\delta)} = g_{+,\delta}(t,\lambda) \, g_{-,\delta}(t,\lambda)\ , \quad \lambda \in i\boldsymbol{R}\ , \tag{4.24}$$

(we take the principal branch of the logarithm) with $g_{\pm,\delta}(t,\cdot) \in e^{\underline{g}_\pm}$ and

$$g_{+,\delta}(t,\infty) = g_{-,\delta}(t,\infty) = (C^2 + \delta)^{t/2}\ . \tag{4.25}$$

From (2.111),

$$g_{+,\delta}(t,\lambda) = g_{-,\delta}(t,-\bar{\lambda})^*\ , \quad \text{Re}\ \lambda > 0\ , \tag{4.26}$$

and inserting the polar decomposition

$$g_{+,\delta}(t,\lambda) = |g_{+,\delta}(t,\lambda)| \, Q_{+,\delta}(t,\lambda)$$

in (4.23) we find

$$|g_{+,\delta}(t,\lambda)| = e^{(t/2)\log(A(\lambda)+\delta)}\ , \tag{4.27}$$

which shows in particular that $|g_{+,\delta}(\lambda,t)|$ is uniformly bonded for $\lambda \in i\boldsymbol{R}$, $1 > \delta > 0$ and $t > 0$. The same is then true for $g_{+,\delta}(t,\lambda)$ by (4.26) as $Q_{+,\delta}(t,\lambda)$ is unitary, and hence

$$\begin{aligned} &g_{\pm,\delta}(t,\lambda) \text{ are uniformly bounded for } t > 0,\ 1 > \delta > 0\ , \\ &\text{and } \operatorname{Re}\lambda \geq 0,\ \ \operatorname{Re}\lambda \leq 0 \text{ respectively,} \end{aligned} \tag{4.28}$$

by the maximum principle.

Next observe that as $\det g_{+,\delta}(t,\lambda)$ is analytic in $\operatorname{Re}\lambda > 0$, it is determined by its absolute value on $i\boldsymbol{R}$. Indeed taking determinants in (4.27) and using Cauchy's formula, we find

$$\det g_{+,\delta}(t,\lambda) = \overline{\det g_{-,\delta}(t,-\bar{\lambda})} = \det(C^2+\delta)^{1/2}\, e^{t\int_{i\infty}^{-i\infty}\log[\frac{\det\,(A(s)+\delta)}{\det\,(C^2+\delta)}]\frac{\bar{d}s}{s-\lambda}} . \tag{4.29}$$

Let

$$A(\lambda) = u(\lambda)\begin{pmatrix} \mu_1(\lambda) & & O \\ & \ddots & \\ O & & \mu_N(\lambda) \end{pmatrix} u(\lambda)^*\ , \quad \lambda \in i\boldsymbol{R}\ , \tag{4.30}$$

denote the (real analytic) spectral decomposition of $A(\lambda)$. From the formula

$$\begin{aligned} g_{+,\delta}(t,\lambda) &= e^{t\log(A(\lambda)+\delta)}\,(g_{-,\delta}(t,\lambda))^{-1} \\ &= u(\lambda)\begin{pmatrix} (\mu_1(\lambda)+\delta)^t & & O \\ & \ddots & \\ O & & (\mu_N(\lambda)+\delta)^t \end{pmatrix} u^*(-\bar{\lambda})\frac{\operatorname{adj}(g_{-,\delta}(t,\lambda))}{\det(g_{-,\delta}(t,\lambda))}\ , \end{aligned} \tag{4.31}$$

we see that for each $\delta > 0$, $g_{+,\delta}(\lambda,t)$ has an analytic continuation in a neighborhood of $i\boldsymbol{R}$, including $\pm i\infty$ (to see this use the $z = (\lambda-1)/(\lambda+1)$ variable of Section 2). Moreover using the fact that the zero eigenvalues of $A(\lambda)$ vanish algebraically to fixed order, in fact precisely to second order (see Lemma 4.16), at $\lambda = 0$, we see that $g_{+,\delta}(\lambda,t)$ has an analytic continuation into a *fixed* region of the shape indicated in Fig. 4.2,

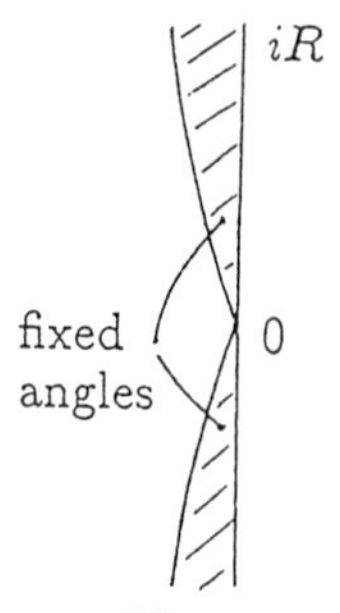


Figure 4.2

which is bounded uniformly on compact subsets (including $\lambda = \infty$) of this region for all $0 < \delta < 1$ and $0 < t < T$ for any $T < \infty$. It follows that there exists $\delta_n \downarrow 0$ and a function $g_+(\lambda, t)$ analytic in this fixed region such that

$$g_+(\lambda, t) = \lim_{n \to \infty} g_{+,\delta_n}(\lambda, t) \ , \tag{4.32$_+$}$$

and similarly there exists $g_-(\lambda, t)$ analytic in the reflected region such that

$$g_-(\lambda, t) = \lim_{n \to \infty} g_{-,\delta_n}(\lambda, t) \ . \tag{4.32$_-$}$$

A priori, there are many such limit functions. From the calculations that follow we will see that they are all equal and hence that

$$\lim_{\delta \downarrow 0} g_{\pm,\delta}(\lambda, t)$$

exist.

Necessarily,

$$\det g_+(\lambda, t) = \overline{\det g_-(-\bar{\lambda}, t)} = \det C'^t e^{t \int_{i\infty}^{-i\infty} \log\left(\frac{\det A(s)}{\det C^2}\right) \frac{ds}{s-\lambda}} \tag{4.33}$$

and we see in particular that $\det g_\pm(\lambda, t) \neq 0$ in $\{\text{Re } \lambda \geq 0\} \setminus 0$, $\{\text{Re } \lambda \leq 0\} \setminus 0$ respectively. Moreover, direct evaluation of (4.33) shows that $\det g_\pm(\lambda, t)$ satisfy (4.20): the details are left as an exercise to the reader.

Conditions (ii), (iii) and (iv) of Theorem 4.18 have now been proved. Conditions (i) and (v) follow by letting $\delta_n \downarrow 0$ in (4.25) and (4.24) respectively (recall that the convergence of $g_{\pm,\delta_n}(t,\lambda)$ is uniform in a neighborhood of $\infty$). It remains to prove uniqueness. So suppose $g'_+, g'_-$ is another factorization with the same properties. Then

$$h \equiv (g'_+)^{-1} g_+ = g'_-(g_-)^{-1} \quad \text{on } iR$$

has an analytic continuation to $\mathbb{C} \setminus 0$. But from (iii) and (iv), $h(\lambda)\lambda^q$ is bounded in a neighborhood of $\lambda = 0$ for some integer $q$. Hence the singularity at $\lambda = 0$ is removable and so, using (v),

$$\lambda^q h(\lambda) = B_0 + B_1\lambda + \cdots + B_q\lambda^q\,, \quad \lambda \in \mathbb{C} \setminus 0\,,$$

and hence

$$g_+(\lambda) = g'_+(\lambda)\Big(\frac{B_0}{\lambda^q} + \frac{B_1}{\lambda^{q-1}} + \cdots + B_q\Big)\,, \quad \lambda \in iR \setminus 0\,, \tag{4.34}$$

for suitable matrices $B_0, B_1, \ldots, B_q$. But for the polar decompositions $g_+ = |g_+|Q_+$, $g'_+ = |g'_+|Q'_+$,

$$|g_+(\lambda)| = |g'_+(\lambda)| = e^{(t/2)\log A(\lambda)}\,, \quad \lambda \in iR \setminus 0\,,$$

and we obtain

$$Q_+(\lambda) = Q'_+(\lambda)\Big(\frac{B_0}{\lambda^q} + \frac{B_1}{\lambda^{q-1}} + \cdots + B_q\Big)\,, \qquad \lambda \in iR \setminus 0\,.$$

In particular

$$\frac{B_0}{\lambda^q} + \frac{B_1}{\lambda^{q-1}} + \cdots + B_q = (Q'_+(\lambda))^* \, Q_+(\lambda)\,, \qquad \lambda \in iR \setminus 0$$

and letting $\lambda \to 0$, we find $B_0 = B_1 = \cdots = B_{q-1} = 0$, as the unitary matrices $Q_+, Q'_+$ are bounded. On the other hand, letting $\lambda \to \infty$ in (4.34), and using (v) again, we obtain $B_q = I$. Thus $h(\lambda) \equiv I$ and we are done. This completes the proof of Theorem 4.18.

**Remark.** The above calculations clearly show that $\lim_{\delta\downarrow 0}$ must exist, and give the above factorization.

Our next goal is to show that

$$A(t,\lambda) = g_+(t,\lambda)^{-1} A(\lambda) g_+(t,\lambda)$$

induces an interpolating flow for the billiard ball map. Because of the singularity at $\lambda = 0$, we must proceed by a limit procedure. The final result is Theorem 4.48 below. So consider

$$A_\delta(t,\lambda) = g_{+,\delta}(t,\lambda)^{-1} A(\lambda) g_{+,\delta}(t,\lambda) \tag{4.35}$$

where $g_{+,\delta}(t,\lambda)$ is given by (4.24),(4.25). Arguing as in Section 2 we find

$$A_\delta = \frac{B_0 + \lambda B_1 - \lambda^2 C^2}{1-\lambda^2} \tag{4.36}$$

for suitable matrices $B_0$ and $B_1$. From (4.27) we also have the formula

$$\begin{aligned} A_\delta(t,\lambda) &= Q^*_{+,\delta}(\lambda) A Q_{+,\delta}(\lambda) \\ &= \frac{Q^*_{+,\delta}(\lambda)y \otimes Q^T_{+,\delta}(\lambda)y + \lambda Q^*_{+,\delta}(\lambda)(x\wedge y)Q_{+,\delta}(\lambda) - \lambda^2 Q^*_{+,\delta}(\lambda)C^2 Q_{+,\delta}(\lambda)}{1-\lambda^2}, \end{aligned}$$

which shows that

$$B_0 = Q^*_{+,\delta}(0)y \otimes Q^T_{+,\delta}(0)y \ . \tag{4.37}$$

As $A(\lambda) = \overline{A(\bar\lambda)}$, we must have as before,

$$g_{\pm,\delta}(t,\lambda) = \overline{g_{\pm,\delta}(t,\bar\lambda)} \ , \tag{4.38}$$

and hence

$$Q_{+,\delta}(\lambda) = \overline{Q_{+,\delta}(\bar\lambda)} \ , \qquad \lambda \in iR \ , \tag{4.39}$$

and, in particular, we see that $Q_{+,\delta}(0)$ is real. Thus

$$B_0 = y_\delta \otimes y_\delta \tag{4.40}$$

where

$$y_\delta = Q^T_{+,\delta}(0)y \ , \qquad \|y_\delta\| = 1 \ . \tag{4.41}$$

Differentiation at $\lambda = 0$ yields similarly

$$B_1 = \tilde{x}_\delta \wedge y_\delta \tag{4.42}$$

where

$$\tilde{x}_\delta = Q^T_{+,\delta}(0,t)x - \frac{d}{d\lambda}\bigg|_{\lambda=0} Q^T_{+,\delta}(\lambda,t)y \ , \tag{4.43}$$

which is real by (4.39).

Assembling these formulae we have

$$A_\delta(t,\lambda) = (y_\delta \otimes y_\delta + \lambda\, \tilde{x}_\delta \wedge y_\delta - \lambda^2 C^2)/(1-\lambda^2)$$

where $\tilde{x}_\delta, y_\delta$ are real and $\|y_\delta\| = 1$. We now show that the flow projects uniquely

$$(\tilde{x}_\delta, y_\delta) \to (x_\delta, y_\delta)$$

to a flow on $Y_+$. To do this note first from (4.35) that

$$\det A_\delta(t,\lambda) = \det A(\lambda)$$

which implies by the Weinstein-Aronszajn formulae that

$$\begin{aligned}(-1)^N \lambda^{2N-2}(\lambda^2 - [(y_\delta, C^{-2}\tilde{x}_\delta)^2 + (1 - \|C^{-1}\tilde{x}_\delta\|^2)\|C^{-1}y_\delta\|^2]) \\ = (-1)^N \lambda^{2N-2}(\lambda^2 - (y, C^{-2}x)^2) \ .\end{aligned} \tag{4.44}$$

On the other hand $A_\delta(t,\lambda)$ is clearly invariant under the translation of $\tilde{x}_\delta$ by a multiple of $y_\delta$,

$$\tilde{x}_\delta \to x_\delta = \tilde{x}_\delta + s\, y_\delta \ .$$

We choose $s \in \boldsymbol{R}$ such that

$$(y_\delta, C^{-2}x_\delta) = (y, C^{-2}x) > 0 \ . \tag{4.45}$$

Then $x_\delta$ automatically lies on $\partial E$ by (4.44). To achieve (4.45) choose

$$s = \frac{(x, C^{-2}y) - (\tilde{x}_\delta, C^{-2}y_\delta)}{\|C^{-2}y_\delta\|^2} \ . \tag{4.46}$$

Thus the flow projects uniquely onto $Y_+$,

$$A_\delta(t,\lambda) = \frac{y_\delta \otimes y_\delta + \lambda x_\delta \wedge y_\delta - \lambda^2 C^2}{1-\lambda^2}, \quad (x_\delta, y_\delta) \in Y_+ . \tag{4.47}$$

From (4.47) we obtain

$$y_\delta \otimes y_\delta = \lambda^2 C^2 + \frac{1-\lambda^2}{2}\left[g_{+,\delta}^{-1}(\lambda)\, A(\lambda)\, g_{+,\delta}(\lambda) + g_{+,\delta}(-\lambda)^{-1}\, A(-\lambda)\, g_{+,\delta}(-\lambda)\right],$$

with a similar formula for $x_\delta \wedge y_\delta$. Letting $\delta \downarrow 0$ for fixed $\lambda \in iR \setminus 0$, the right hand side converges: hence

$$y_\delta \otimes y_\delta \to Z ,$$

and similarly

$$x_\delta \wedge y_\delta \to W ,$$

for suitable $Z, W$. For $\delta > 0$, we have $\|y_\delta\| = 1 = \|C^{-2}x_\delta\|$, $(x_\delta, C^{-2}y_\delta) = (x, C^{-2}y)$. Suppose $y_{\delta_n} \to \hat{y}$, $x_{\delta_n} \to \hat{x}$ for some sequence $\delta_n \downarrow 0$. Then $\|\hat{y}\| = 1 = \|C^{-1}\hat{x}\|$, $(\hat{x}, C^{-2}\hat{y}) = (x, C^{-2}y)$, and also

$$\hat{y} \otimes \hat{y} = Z , \qquad \hat{x} \wedge \hat{y} = W .$$

Now suppose $y_{\delta'_n} \to \hat{y}'$ for some other sequence $\delta'_n \downarrow 0$. Then the above calculations show that $\hat{y}' = \pm\hat{y}$. But we must have $\hat{y}' = \hat{y}$: otherwise, by the geometry of the unit sphere, there exists a sequence $\delta''_n \downarrow 0$ with $y_{\delta''_n}$ perpendicular to $\hat{y}$ and $y_{\delta''_n} \to \hat{y}''$. This implies

$$\hat{y}'' \otimes \hat{y}'' = Z = \hat{y} \otimes \hat{y} ,$$

which is a contradiction. We conclude that

$$y(t) = \lim_{\delta \downarrow 0} y_\delta(t)$$

exists. As $\hat{x}$ is uniquely determined from $W = \hat{x} \wedge \hat{y} = \hat{x} \wedge y(t)$, $(\hat{x}, C^{-2}\hat{y}) = (\hat{x}, C^{-2}y(t)) = (x, C^{-2}y)$, we also conclude that

$$x(t) = \lim_{\delta \downarrow 0} x_\delta(t)$$

exists. Necessarily $(x(t), y(t)) \in Y_+$ with $(x(t), C^{-2}y(t)) = (x, C^{-2}y)$.

**Theorem 4.48.** *If*

$$e^{t \log A)(\lambda)} = g_+(t,\lambda)\, g_-(t,\lambda)$$

*is the factorization given in Theorem 4.18, then*

$$A(t,\lambda) = g_+(t,\lambda)^{-1}\, A(\lambda)\, g_+(t,\lambda)$$

*has the form*

$$A(t,\lambda) = \frac{y(t) \otimes y(t) + \lambda\, x(t) \wedge y(t) - \lambda^2 C^2}{1-\lambda^2}\,, \tag{4.49}$$

*where $(x(t), y(t)) \in C^1([0,\infty), Y_+)$, and solves the flow*

$$\begin{aligned} \frac{d}{dt} A(t,\lambda) &= \left[(\pi_- \log A(t,\cdot))(\lambda),\ A(t,\lambda)\right]\,, \\ A(0,\lambda) &= A(\lambda)\,, \end{aligned} \tag{4.50}$$

*uniquely. Furthermore the flow*

$$(x,y) \to (x(t), y(t))$$

*induced on $Y_+$ interpolates the skew-hodograph mapping $\phi$ at integer times.*

**Proof:** The fact that $(x(t), y(t))$ is $C^1$ and that $A(t,\lambda)$ solves (4.50), follows by a bootstrap argument, letting $\delta \downarrow 0$ in the integral equation

$$A_\delta(t,\lambda) = A(\lambda) + \int_0^t \left[(\pi_- \log(A_\delta(s,\cdot) + \delta))(\lambda)\ ,\ A_\delta(s,\lambda)\right] ds\ .$$

We leave the details to the reader. But (4.50) implies in turn that

$$V(x,y,\lambda) \equiv \left[(\pi_- \log A(x,y,\cdot))(\lambda)\ ,\ A(x,y,\lambda)\right]$$

is a vector field on matrices of the form

$$A(x,y,\lambda) = (y \otimes y + \lambda\, x \wedge y - \lambda^2 C^2)/(1-\lambda^2)\,, \quad (x,y) \in Y_+\,,$$

and hence induces a vector field on $Y_+$. To prove uniqueness for the solution of (4.50), it is enough to prove that $V(x, y, \lambda)$ is smooth in the variables $x$ and $y$. This is an exercise in complex variables using the properties of the roots $\mu(\lambda)$ in Lemma 4.14. Again the details are left to the reader.

Finally to prove that the flow

$$(x(t), y(t)) = (x(t; x_0, y_0)\ ,\ y(t; x_0, y_0))$$

interpolates $\phi$, recall (4.22) which implies

$$\begin{aligned} &(x(t; x_0, y_0)\ ,\ y(x; x_0, y_0)) = \chi(x_0, y_0)\,\phi(x_0, y_0) \\ &\text{where} \\ &\chi(x_0, y_0) = \pm 1\ . \end{aligned} \tag{4.51}$$

But as $\phi(x_0, y_0)$ and $(x(t; x_0, y_0), y(t; x_0, y_0))$ depend continuously on $(x_0, y_0)$ (note $V(x, y, \lambda)$ is smooth in $x$ and $y$), $\chi(x_0, y_0)$ is also continuous, and hence it is sufficient to prove $\chi(x_0, y_0) = +1$ for one point $(x_0, y_0) \in Y_+$. We take $(x_0, y_0) = (Ce_N, e_N)$, where $e_N = (0, \ldots, 0, 1)^T$. In this case,

$$A(\lambda) = (e_N \otimes e_N - \lambda^2 C^2)/(1 - \lambda^2)\ ,$$

which is diagonal. Hence $\log A(\lambda)$ is diagonal, which implies that $g_\pm(t, \lambda)$ are diagonal, and so

$$A(t, \lambda) = g_+(t, \lambda)^{-1}\, A(\lambda)\, g_+(t, \lambda) = A(\lambda) = \text{constant}\ .$$

In particular

$$(x(t; x_0, y_0)\ ,\ y(t; x_0, y_0)) = (x_0, y_0)\ .$$

But clearly from the geometry of $E$,

$$\Psi_k(x_0, y_0) = (-x_0, -y_0)\ ,$$

and we obtain from (4.5)

$$\phi(x_0, y_0) = (-C(-y_0)\ ,\ C^{-1}x_0)$$
$$= (x_0, y_0)\ ,$$

and we are done. □

**(b) The interpolating flow as a constrained motion**

In Section 2 we showed that the induced flow $t \to M(t)$ was Hamiltonian and integrable on $o_2(N)^*$. In this section we prove a similar result for the induced flow $t \to (x(t), y(t))$ on $Y_+$. It is well known (see [B]), that in the case $N = 2$ the billiard map $\Psi_b$ is symplectic with respect to the Birkhoff 2-form $\omega_B$ (see (4.54) below) on $Y_+$. In $N$ dimensions we will use an appropriate generalization of the Birkhoff form.

First we introduce the Birkhoff variables $(\theta, \psi)$ for $Y_+$ in the case $N = 2$. Let $C = \operatorname{diag}(c_1, c_2)$.

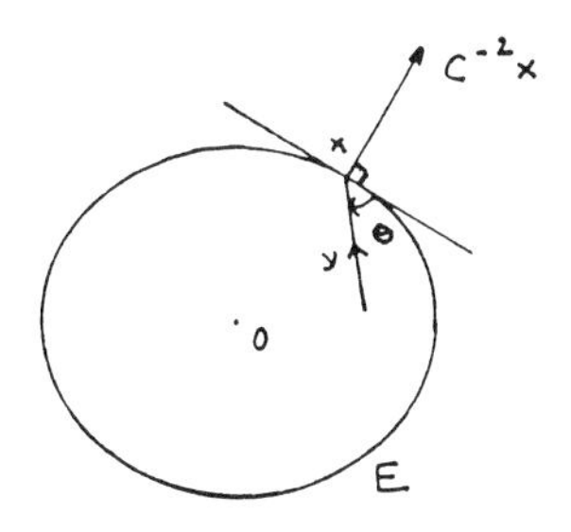


Figure 4.3

We have

$$Y_+ \cong \{(\theta, \psi) : 0 < \theta < \pi\ ,\ \psi \in R \setminus 2\pi\mathbb{Z}\}\ ,$$

$$x = (c_1 \cos\psi\ ,\ c_2 \sin\psi) \tag{4.52}$$
$$y = \left(\frac{\cos\psi \sin\theta}{c_1} - \frac{\sin\psi\cos\theta}{c_2}\ ,\ \frac{\sin\psi\sin\theta}{c_2} + \frac{\cos\psi\cos\theta}{c_1}\right) / \|C^{-2}x\|\ .$$

We find

$$(x, C^{-2}y) = \|C^{-2}x\| \sin\theta\ , \tag{4.53}$$

and the Birkhoff 2-form is given by

$$\omega_B = c_1c_2 \sin\theta \|C^{-2}x\|\, d\theta \wedge d\psi = c_1c_2(x, C^{-2}y)\, d\theta \wedge d\psi \ . \tag{4.54}$$

We recall some facts about the restriction of 2-forms $\omega$ on a manifold $V$ to an even-dimensional submanifold $Z$, which is presented in the form

$$Z = \{v \in V : f_i(v) = 0\ , \quad 1 \le i \le 2q\} \tag{4.55}$$

as the intersection of the zero level sets of $2q$ smooth and independent functions $f_1, \ldots, f_{2q}$ (see, for example, [M] or [DLTr]). The factors are these.

(i) $\omega_Z \equiv \omega\big|_{T(Z)\times T(Z)}$ is a nondegenerate 2-form on $Z$ if and only if the matrix of Poisson brackets $(\{f_i, f_j\})_{1\le i,j\le 2q}$ is invertible.

(ii) Smooth functions $F$ on $V$ give rise to Hamiltonians on $(Z, \omega_Z)$ simply by restriction. If $\{\cdot,\cdot\}$ denotes the Poisson structure on $Z$ associated with $\omega_Z$, then

$$\{F_1, F_2\}_Z = \{F_1, F_2\} \tag{4.56}$$

provided

$$\{F_i, f_j\} = 0\ , \qquad 1 \le j \le 2q$$

for $i = 1$ *or* 2. This suggests a convenient way to compute the flow induced on $(Z, \omega_Z)$ by a Hamiltonian $H$ defined on the full space $V$: choose scalars $\beta_1, \ldots, \beta_{2q}$ such that $\{H + \sum_{i=1}^{2q} \beta_i f_i,\ f_j\} = 0$ for $j = 1, \ldots, 2q$. Then

$$\begin{aligned}\{H, F\}_Z &= \{H + \sum_{i=1}^{2q} \beta_i f_i\ ,\ F\}_Z \\ &= \{H + \sum_{i=1}^{2q} \beta_i f_i\ ,\ F\}\ , \qquad \text{by (4.56)},\end{aligned}$$

which shows that all computations can be carried in the ambient space $V$.

**Lemma 4.57.** *Let $\omega = \sum_{i=1}^N dx_i \wedge dy_i$ be the standard 2-form on $\mathbf{R}^{2N}$. Then the restriction $\omega_{Y_+}$ of $\omega$ to $Y_+$ is nondegenerate. Moreover, in the case $N = 2$, $\omega_{Y_+} = \omega_B$.*

**Proof:** To verify the nondegeneracy, observe that

$$\{(x, C^{-2}x)\ ,\ (y, y)\} = 4(x, C^{-2}y)\ ,$$

which is nonzero on $Y_+$.

To verify $\omega_{Y_+} = \omega_B$ in the case $N = 2$, simply insert the formulae in (4.52) for $x$ and $y$, into the expression $dx_1 \wedge dy_1 + dx_2 \wedge dy_2$. □

For $(x, y) \in R^{2N} \setminus \{x = 0 \text{ or } y = 0\}$, consider the Hamiltonian

$$H(x, y) = \frac{1}{2} \int_{C_R} \left[(\mathrm{tr}(\hat{A}(\lambda) \log \hat{A}(\lambda) - \hat{A}(\lambda))) + 1\right] \left(\frac{1 - \lambda^2}{\lambda^2}\right) d\lambda\ . \tag{4.58}$$

Here $C_R$ is the contour

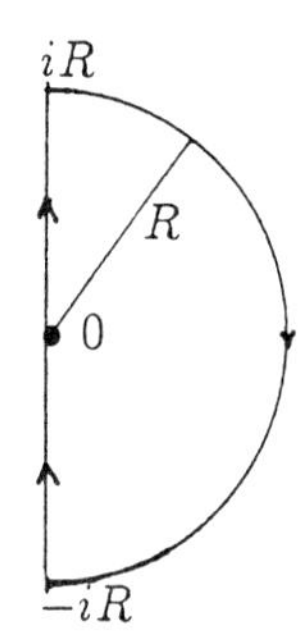


Figure 4.4

for any sufficiently large $R$, and

$$\widehat{A}(\lambda) = \frac{\hat{y} \otimes \hat{y} + \lambda \hat{x} \wedge \hat{y} - \lambda^2 C^2}{1 - \lambda^2}\ , \tag{4.59}$$

where

$$\hat{x} = x / \|C^{-1}x\|\ , \qquad \hat{y} = y / \|y\|\ . \tag{4.60}$$

Clearly $(\hat{x}, \hat{y}) \in Y$. Again we choose the principal branch of the logarithm. By Cauchy's theorem, $H$ is clearly independent of $R$, as long as $R$ is sufficiently large.

The function $H$ is well-defined by the following computation. $\widehat{A}(0)$ has one eigenvalue $\mu_1(0) = 1$, and $N - 1$ eigenvalues $\mu_j(0) = 0$, $j = 2, \ldots, N$. By Lemma 4.14, $\mu_j(\lambda) \sim \lambda^2$

as $\lambda \to 0$, $j = 2, \ldots, N$. But $\sum_{i=1}^{N} \mu_i(\lambda) = \operatorname{tr} \widehat{A}(\lambda) = (1 - \lambda^2 \operatorname{tr} C^2)/(1-\lambda^2)$, and hence $\mu_1(\lambda) = 1 + O(\lambda^2)$ as $\lambda \to 0$. Thus

$$\lambda^{-2}\left[\operatorname{tr}(\widehat{A}(\lambda)\log \widehat{A}(\lambda) - \widehat{A}(\lambda)) + 1\right] = \sum_{i=2}^{N} \frac{\mu_i \log \mu_i - \mu_i}{\lambda^2} + \frac{\mu_1 \log \mu_1 - \mu_1 + 1}{\lambda^2},$$

which is in $L^1(d\lambda)$ near $\lambda = 0$, by the above estimates on the $\mu_i$'s.

**Theorem 4.61.** *The flow generated by $H$ on $(Y_+, \omega_{Y_+})$ coincides precisely with the induced interpolating flow of Theorem* 4.48.

**Proof:** On $Y_+$ we drop the $\hat{}$ notation. We have for Re $\lambda < 0$,

$$\begin{aligned}
&\left[\pi_- \log A(\cdot), A\right](\lambda) \\
&= \int \left[\log A(\lambda'), \frac{A(\lambda) - A(\lambda')}{\lambda - \lambda'}\right] d\lambda' \\
&= \frac{1}{1-\lambda^2} \int \left[\log A(\lambda'), (\lambda' C^2 - x \wedge y - \lambda' y \otimes y) + \lambda(C^2 - y \otimes y - \lambda' x \wedge y)\right] \frac{đ\lambda'}{1 - \lambda'^2},
\end{aligned}$$

which implies by (4.50) that

$$\begin{aligned}
\frac{d}{dt} y \otimes y &= \int \left[\log A(\lambda'), \frac{1}{\lambda'}(\lambda'^2 C^2 - \lambda' x \wedge y - y \otimes y) + ((\lambda')^{-1} - \lambda') y \otimes y\right] \frac{đ\lambda'}{1 - \lambda'^2} \\
&= \int \left[\log A(\lambda'), y \otimes y\right] \frac{đ\lambda'}{\lambda'},
\end{aligned} \tag{4.62}$$

and similarly

$$\frac{d}{dt} x \wedge y = \int \left[\log A(\lambda'), C^2\right] đ\lambda'. \tag{4.63}$$

Now consider the Hamiltonian

$$H' = H + \eta((x, C^{-2}x) - 1) + \nu(\|y\|^2 - 1) \tag{4.64}$$

on $\boldsymbol{R}^{2N}$. We will choose the scalar $\eta, \nu$ such that $\{H', (x, C^{-2}x)\} = \{H', \|y\|^2\} = 0$ (see (ii) above).

The condition $\{H', \|y\|^2\} = 0$ implies that

$$\eta = -\{H, \|y\|^2\} / 4(x, C^{-2}y). \tag{4.65}$$

But on $Y_+$, we find

$$\frac{\partial H}{\partial x_i} = \frac{1}{2}\int_{C_R} \left(\operatorname{tr}\left[\log A(\lambda)e_i \wedge y)\right] - \operatorname{tr}\left[\log A(\lambda)(x \wedge y)\right](C^{-2}x)_i\right)\frac{đ\lambda}{\lambda} \tag{4.66}$$

($e_i$ is the standard basis vector) and so

$$\{H, \|y\|^2\} = 2\sum_i \frac{\partial H}{\partial x_i} y_i = \int_{C_R} \left(\operatorname{tr}\left[(\log A(\lambda))\, y \wedge y\right] - \operatorname{tr}\left[(\log A(\lambda))\, x \wedge y\right](y, C^{-2}x)\right)\frac{đ\lambda}{\lambda}$$

which gives

$$\eta = \frac{1}{4}\int_{C_R} \operatorname{tr}\left[(\log A(\lambda))(x \wedge y)\right]\frac{đ\lambda}{\lambda}\,. \tag{4.67}$$

Inserting (4.66) and (4.67) in

$$\frac{dy_j}{dt} = -\frac{\partial H'}{\partial x_j} = -\frac{\partial H}{\partial x_j} - \eta\frac{2x_j}{c_j^2}\,,$$

we find

$$\begin{aligned}\frac{dy_i}{dt} &= -\frac{1}{2}\int_{C_R} \operatorname{tr}\left((\log A)(\lambda)(e_j \wedge y)\right)\frac{đ\lambda}{\lambda} \\ &= -\frac{1}{2}\int_{-i\infty}^{i\infty} \operatorname{tr}\left((\log A(\lambda))(e_j \wedge y)\right)\frac{đ\lambda}{\lambda}\,,\end{aligned}$$

(the contribution from the semicircle vanishes as $\log A(\lambda) \to \log C^2$ and $e_j \wedge y$ is skew)

$$= \frac{1}{2}\int_{-i\infty}^{i\infty} \left((\log A)(\lambda)\, y\right)_j \frac{đ\lambda}{\lambda} - \frac{1}{2}\int_{-i\infty}^{i\infty} \left((\log A)(-\lambda) y\right)_j \frac{đ\lambda}{\lambda}$$

as $A^T(\lambda) = A(-\lambda)$. Thus

$$\frac{dy}{dt} = \left(\int \log A(\lambda)\frac{đ\lambda}{\lambda}\right) y\,. \tag{4.68}$$

On the other hand from (4.62)

$$\begin{aligned}\frac{dy}{dt} &= \int \left[(\log A(\lambda))y - (y, \log A(\lambda)y)y\right]\frac{đ\lambda}{\lambda} \\ &= \left(\int \log A(\lambda)\frac{đ\lambda}{\lambda}\right)y\,, \quad ((y, \log A(\lambda)y) \text{ is even})\end{aligned}$$

which agrees with (4.68).

To compute $dx/dt$, we first determine $\nu$ from the requirement $\{H', (x, C^{-2}x)\} = 0$, and then substitute in

$$\frac{dx_i}{dt} = \frac{\partial H}{\partial y_i} + \nu(2y_i) ,$$

which leads to the formula

$$\frac{d}{dt} x \wedge y = \int (\log A(\lambda) y) \wedge y \frac{đ\lambda}{\lambda^2} + \int (\log A(\lambda) x) \wedge y \frac{đ\lambda}{\lambda} + \int x \wedge (\log A(\lambda) y) \frac{đ\lambda}{\lambda} . \quad (4.69)$$

On the other hand from (4.63),

$$\int \left[\log A(\lambda), C^2\right] đ\lambda = \int \left[\log A(\lambda) \, , x \wedge y\right] \frac{đ\lambda}{\lambda} + \int \left[\log A(\lambda) \, , y \otimes y\right] \frac{đ\lambda}{\lambda^2} .$$

Writing out the terms, and again using $A(\lambda)^T = A(-\lambda)$, we easily obtain the right hand side of (4.69). This completes the proof of the theorem. □

The above calculations do not rely on any special properties of the logarithm. If we take the Hamiltonian

$$H_F = \frac{1}{2} \int_{C_R} F(A(\lambda)) \frac{đ\lambda^2}{\lambda^2 (1 - \lambda^2)^{-1}}$$

for any $F$ which is real analytic on $(0, \infty)$ (provided the integral exists), then the above calculations show that the flow generated by $H_F$ on $(Y_+, \omega_{Y_+})$, coincides with the flow

$$\frac{dA}{dt} = \left[\pi_- F'(A), A\right]$$

on $\underline{\tilde{g}}^*_{\text{sing}}$, where again $A = (y \otimes y + \lambda x \wedge y - \lambda^2 C^2)/(1 - \lambda^2)$. Now all these flows commute on $\underline{\tilde{g}}^*_{\text{sing}}$, as well as providing integrals for the billiard flow. It follows that they commute on $Y_+$. If we take

$$I_j = \int_{C_R} (\text{tr } A(\lambda)^j - \text{tr } A(0)^j) \frac{đ\lambda}{\lambda^2 (1 - \lambda^2)^{-1}} , \quad 2 \le j \le N ,$$

we obtain $N - 1$ commuting integrals on $Y_+$. In particular *the induced interpolating flow of Theorem* 4.48 *is completely integrable.* Of course the integrals are equivalent to the eigenvalues of $y \otimes y + \lambda x \wedge y - \lambda^2 C^2$, which are in turn equivalent to the coefficients of the curve $\det(y \otimes y + \lambda x \wedge y - \lambda^2 C^2 - z) = 0$.

**Remark 4.70.** From Theorem 4.48 and formula (4.6) we see that two steps of the generalized Cholesky flow on $\underline{\tilde{g}}^*$ give the billiard map (up to a sign). As is well known, however, two steps of the classical Cholesky algorithm correspond to one step of QR (see, for example, [Wi]). A generalized QR algorithm on $\underline{\tilde{g}}^*$ should then naturally give the billiard map (up to a sign) after one step.

We consider a factorization

$$e^{t \log A(\lambda)} = Q(\lambda)\, g_-(\lambda)\ , \quad \lambda \in iR \tag{4.71}$$

where $g_-(\lambda)$ is analytic in Re $\lambda < 0$, $g_-(\infty)$ is positive and diagonal, and $Q(\lambda)$ is unitary on $iR$. Note that $Q(\infty) = e^{t \log C^2}\, g_-(\infty)^{-1}$ is unitary, positive and diagonal, and hence $Q(\infty)$ necessarily equals $I$. Also

$$\begin{aligned} e^{2t \log A(\lambda)} &= (e^{t \log A(\lambda)})^*\, e^{t \log A(\lambda)} \\ &= g_-(\lambda)^*\, g_-(\lambda) \\ &= g_-(-\bar{\lambda})^*\, g_-(\lambda) \end{aligned}$$

and we see that $g_-(t,\lambda) = g_-^{Ch}(2t,\lambda)$, where $g_-^{ch}$ is the Cholesky factor in Theorem 4.18. Thus the generalized QR flow,

$$\begin{aligned} A^{QR}(t,\lambda) &= g_-(t,\lambda)\, A(\lambda)\, g_-(t,\lambda)^{-1} \\ &= g_-^{Ch}(2t,\lambda)\, A(\lambda)\, g_-^{ch}(2t,\lambda)^{-1} \end{aligned}$$

interpolates the billiard map (up to a sign) at integer times, as desired.

The generalized QR flow and the factorization (4.71) can be interpreted in terms of a loop group framework. In terms of the $z$ variable for the Lie algebra $\underline{g}$ of Section 2, set

$$\begin{aligned} (\pi_k X)(z) &= X_+(z) - X_+(z)^* + X_+(1)^* - X_+(1)\ , \\ (\pi_\ell X)(z) &= X_-(z) + X_+(z)^* + X_0 + X_+(1) - X_+(1)^*\ . \end{aligned}$$

Then $R^{QR} = \pi_k - \pi_\ell$ is the desired $R$-matrix on $\underline{\tilde{g}}$, etc., and, if we so chose, we could proceed to analyze the billiard map using the QR as opposed to the Cholesky framework. Again we would need to consider singular orbits of the form $\underline{\tilde{g}}^*_{\text{sing}}$, etc.

**(c) The interpolating flow in the case $N = 2$.**

In the case $N = 2$, we have $C = \begin{pmatrix} c_1 & 0 \\ 0 & c_2 \end{pmatrix}$, and without loss of generality we can assume $c_1 > c_2 > 0$. A convenient formula for $\log A(\lambda)$ is given by

$$\log A(\lambda) = a(\lambda)\, A(\lambda) + b(\lambda) \tag{4.72}$$

where

$$a(\lambda) = \frac{\log(\mu_1(\lambda)/\mu_2(\lambda))}{\mu_1(\lambda) - \mu_2(\lambda)} \tag{4.73}$$

$$b(\lambda) = \frac{\mu_2(\lambda)\log\mu_1(\lambda) - \mu_1(\lambda)\log\mu_2(\lambda)}{\mu_2(\lambda) - \mu_1(\lambda)} \tag{4.74}$$

and again $\mu_1(\lambda)$, $\mu_2(\lambda)$ are the eigenvalues of $A(\lambda)$. Inserting (4.72) into (4.62) we obtain

$$\begin{aligned} \frac{d}{dt} y \otimes y &= \int [A(\lambda'), y \otimes y]\, a(\lambda') \frac{đ\lambda'}{\lambda'} \\ &= \Big(\int \frac{a(\lambda')}{1 - \lambda'^2} d\lambda'\Big)\, [x \wedge y\, ,\, y \otimes y] - \int [C^2, y \otimes y]\, \frac{\lambda' a(\lambda')}{1 - \lambda'^2}\, đ\lambda' \\ &= \beta\big[x \wedge y,\, y \otimes y\big] \end{aligned} \tag{4.75}$$

where

$$\beta = \int_{-i\infty}^{i\infty} \frac{a(\lambda')}{1 - \lambda'^2}\, đ\lambda' \,, \tag{4.76}$$

and we have used the fact that $A(\lambda)^T = A(-\lambda)$ and hence $a(\lambda)$ is even. Similarly from (4.63),

$$\frac{d}{dt} x \wedge y = \beta\big[y \otimes y,\, C^2\big] \,. \tag{4.77}$$

The (positive) scalar $\beta$ depends only on the eigenvalues of $A = A(t, \lambda')$ and hence is constant under the motion: $\beta$ can be evaluated explicitly in terms of elliptic functions. For example, if the constant of the motion $e = (x, C^{-2} y)$ satisfies

$$0 \le e < (c_1^2 + c_2^2)^{-1/2} \,, \tag{4.78}$$

then we find

$$\beta = \int_0^e \frac{ds}{\sqrt{g(s^2)}} \,, \tag{4.79}$$

where

$$g(u) = \left(1 - u(c_1 + c_2)^2\right)\left(1 - u(c_1 - c_2)^2\right), \tag{4.80}$$

with similar formulae in the remaining cases $(c_1^2 + c_2^2)^{-1/2} \le e \le c_1^{-1}$, $c_1^{-1} < e \le c_2^{-1}$.

In loop group form, (4.75) and (4.77) can be written

$$\frac{d}{dt}(y \otimes y + \lambda x \wedge y - \lambda^2 C^2) = \beta\left[x \wedge y - \lambda C^2, y \otimes y + \lambda x \wedge y - \lambda^2 C^2\right]. \tag{4.81}$$

In terms of the Birkhoff variables $(\theta, \psi)$ a somewhat lengthy but straightforward computation shows that (4.75) and (4.77) take the form

$$\frac{d\theta}{dt} = \frac{\beta(c_1/c_2 - c_2/c_1)}{2e}\sin^2\theta \sin 2\psi \tag{4.82}$$

$$\frac{d\psi}{dt} = -\beta e\, c_1 c_2 \tan\theta, \tag{4.83}$$

which imply in turn,

$$\frac{d^2\psi}{dt^2} + \frac{1}{2}\beta^2(c_2^2 - c_1^2)\sin 2\psi = 0. \tag{4.84}$$

This is the pendulum equation, which can be solved in terms of elliptic functions in the standard way.

It is an amusing and instructive exercise, which we leave to the reader, to carry through the phase space analysis of (4.82) and (4.83) (cf. [G-M]) and connect the results to the geometry of the ellipse $\partial E$. As in the case of the pendulum, there are two types of motion depending on the value of the parameter $e$, one which moves "over the top" and one which does not, with a separatrix at $e = c_1^{-1}$.

Finally we note that the integrability of ellipsoidal billiards is usually described in terms of confocal ellipses

$$\frac{x_1^2}{c_1^2 - \gamma_0} + \frac{x_2^2}{c_2^2 - \gamma_0} = 1 \tag{4.85}$$

as follows: reflection at $\partial E$ is such that the billiard motion in $E$ is always tangential to a fixed confocal ellipse,

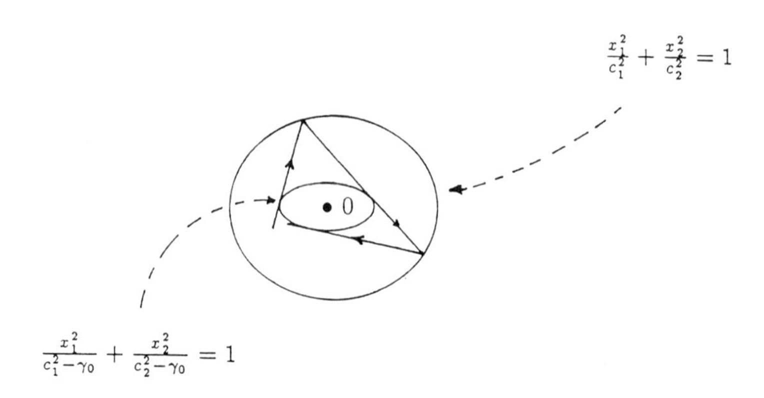


Figure 4.5

It is of interest to express $\gamma_0$ in terms of $e$, and a simple exercise in calculus shows that

$$\gamma_0 = c_1^2 c_2^2 \, e^2 \tag{4.86}$$

In particular at the separatrix $e = c_1^{-1}$, $\gamma_0 = c_2^2$ and the ellipse (4.85) becomes a hyperbola.

## 5. Loop groups and rank 2 extensions

In [M], Moser introduced a class of matrices

$$L = L(x,y) = \Lambda + ax \otimes x + bx \otimes y + cy \otimes x + dy \otimes y \ , \tag{5.1}$$

$$\Delta = ad - bc \neq 0 \ , \qquad x, y \in \boldsymbol{R}^N \ , \tag{5.2}$$

which are rank 2 extensions of a fixed, diagonal matrix

$$\Lambda = \operatorname{diag}(\alpha_1, \ldots, \alpha_N) \ , \qquad \alpha_i \neq \alpha_j \quad \text{for } i \neq j \ , \tag{5.3}$$

and considered flows $L(t) = L(x(t), y(t))$ which are isospectral deformations of $L(t=0)$. This includes a rich and interesting class of dynamical systems, such as the Neumann problem, geodesic flow on an ellipsoid, the Toda lattice, etc. (see [M]).

The basic result in [M] is the following. Let

$$\Phi_z(x,y) \equiv 1 - \frac{\det(z - L(x,y))}{\det(z - \Lambda)} \ . \tag{5.4}$$

By the Weinstein-Aronszajn formulae, one finds

$$\Phi_z(x,y) = a\,Q_z(x) + (b+c)Q_z(x,y) + dQ_z(y) - \Delta(Q_z(x)\,Q_z(y) - Q_z^2(x,y)) \ . \tag{5.5}$$

where

$$Q_z(x,y) = (x, \frac{1}{z - \Lambda} y) \ , \quad Q_z(x) = Q_z(x,x) \ . \tag{5.6}$$

Then the flows generated by the Hamiltonian

$$H = H(x,y) = \frac{1}{4\pi i} \int_{|z|=R} f(z)\,\Phi_z(x,y)\,dz \tag{5.7}$$

(the circle $|z| = R$ runs counterclockwise and contains $\operatorname{spec}(\Lambda)$; $f(z)$ is any polynomial) on $(R^{2N}, \sum_{i=1}^N dx_i \wedge dy_i)$, are isospectral for $L$ and can be written in Lax-pair form

$$\frac{dL}{dt} = [B, L] \ . \tag{5.8}$$

where

$$B = r\beta + \Delta\Gamma\ , \qquad r = (b-c)/2\ , \tag{5.9}$$

and

$$\beta = \operatorname{diag}(\beta_1, \ldots, \beta_N) = \operatorname{diag}(f(\alpha_1), \ldots, f(\alpha_N))\ , \tag{5.10}$$

$$\Gamma = \left(\frac{\beta_i - \beta_j}{\alpha_i - \alpha_j}(x \wedge y)_{ij}\right) = -\Gamma^T\ . \tag{5.11}$$

Furthermore the flows are linearized via the Abel map on the Jacobi variety of the curve

$$\omega^2 = P(z) = \left(r^2 a(z) - \Delta\, \ell(z)\right) a(z)\ , \tag{5.12}$$

where

$$\ell(z) = \det(z - L(x,y)) \quad \text{and} \quad a(z) = \det(z - \Lambda)\ . \tag{5.13}$$

As noted in the Introduction, we will show how these results can be interpreted in terms of a loop group framework, in the following sense.

**Theorem 5.14.** *Associated to every flow* (5.8) *generated by a Hamiltonian of type* (5.7), *there is a loop* $A(x,y,\lambda)$, *with* $(1-\lambda^2)A(x,y,\lambda)$ *quadratic in* $\lambda$, *and a Hamiltonian flow*

$$\frac{d}{dt}A(x(t),y(t),\lambda) = \left[(\pi_- F'(A(x(t),y(t),\cdot)))(\lambda)\ ,\ A(x(t),y(t),\lambda)\right] \tag{5.15}$$

*arising from a factorization of type (1.38) in which the logarithm is replaced by* $F'(\cdot)$, *with the property that, for the appropriate parameter value* $\lambda = \lambda_0$, (5.15) *reduces to* (5.8). *Moreover the curve*

$$\{(\lambda,\eta):\ \det(A(x,y,\lambda) - \eta) = 0\} \tag{5.16}$$

*is precisely the curve* (5.12)*used by Moser in* [M] *to integrate the flow.*

The remainder of this section is devoted to proving this result. First we note with Moser that there are three distinct normal forms for $L$.

(i) $L = \Lambda + a x \otimes x + y \otimes y + r\, x \wedge y$, $a \neq 0$,

or equivalently,

$L = \Lambda + b\, x \otimes y + c\, y \otimes x,$

where

$$s = (b+c)/2 \neq 0 .$$

(ii) $L = \Lambda + dy \otimes y + r\, x \wedge y,\ d \neq 0$

(iii) $L = \Lambda + r\, x \wedge y.$

Cases (ii) and (iii) fit into the loop group framework in a straightforward way. The flow (5.8) becomes

$$\frac{d}{dt}(\Lambda + r\, x \wedge y + d\, y \otimes y) = \left[ r\beta + \Delta\Gamma,\ \Lambda + r\, x \wedge y + d\, y \otimes y \right] . \tag{5.17}$$

But (5.17) is equivalent to

$$\frac{d}{dt}(\lambda^2 \Lambda + \lambda r\, x \wedge y + d\, y \otimes y) = \left[ \lambda r\beta + \Delta\Gamma,\ \lambda^2\Lambda + \lambda r\, x \wedge y + d\, y \otimes y \right] \tag{5.18}$$

for all $\lambda$. Indeed, as $[\beta, \Lambda] = 0$, this is a relation of the form

$$D(\lambda) = D_0 + D_1\lambda + D_2\lambda^2 \equiv 0$$

where the $D_i$ are independent of $\lambda$. As $D(1) = 0$ by (5.17), and as $D(-1) = 0$ by taking transposes, it is enough to show $D(\infty) = 0$: but this reduces to showing

$$\Delta\left[\Gamma, \Lambda\right] + r^2\left[\beta,\ x \wedge y\right] = 0 .$$

In the cases (ii) and (iii), $\Delta = r^2$, and the result follows from the identity

$$[\Gamma, \Lambda] + [\beta,\ x \wedge y] = 0 \tag{5.19}$$

established in [M]. Moreover, by (5.5)

$$\begin{aligned} &\frac{\det(\Lambda + \lambda^{-1} r\, x \wedge y + \lambda^{-2} d\, y \otimes y - z)}{\det(\Lambda - z)} \\ &= 1 - (\lambda^{-2} d\, Q_z(y) - (0 + \lambda^{-2} r^2)(Q_z(x)\, Q_z(y) - Q_z^2(x,y))) \\ &= 1 - \lambda^{-2}\, \Phi_z(x,y) = 1 - \lambda^{-2}(1 - \ell(z)/a(z)) . \end{aligned}$$

Thus if we consider the loop

$$A(\lambda) = \frac{\lambda^2 \Lambda + \lambda r\, x \wedge y + d\, y \otimes y}{\lambda^2 - 1}$$

as in Section 2, then the curve

$$\det(A(\lambda) - \eta) = 0$$

is clearly (birationally) equivalent to

$$\lambda = \sqrt{\frac{a(z) - \ell(z)}{a(z)}} = \frac{\sqrt{(r^2 a(z) - \Delta \ell(z))\, a(z)}}{r\, a(z)} ,$$

or setting $\omega = r\, a(z) \lambda$,

$$\omega^2 = (r^2 a(z) - \Delta \ell(z))\, a(z) ,$$

which is the curve (5.12).

In the case (i), we consider for definiteness the form

$$L = \Lambda + b\, x \otimes y + c\, y \otimes x , \qquad s = \frac{b + c}{2} \neq 0 .$$

The flow (5.8) takes the form

$$\frac{d}{dt}(\Lambda + b\, x \otimes y + c\, y \otimes x) = \left[ r\beta + \Delta\Gamma \,,\, \Lambda + b\, x \otimes y + c\, y \otimes x \right] . \tag{5.20}$$

As we now show, (5.20) is equivalent to

$$\frac{d}{dt} A(\lambda) = \left[ - s\lambda\beta + \Delta\Gamma,\, A(\lambda) \right] \tag{5.21}$$

where

$$A(\lambda) = \Lambda + \frac{\delta}{1 + \lambda} x \otimes y + \frac{\delta}{1 - \lambda} y \otimes x , \tag{5.22}$$

and

$$\delta = 2bc/(b + c) = -\Delta / s . \tag{5.23}$$

As before, this is a relation of the form

$$D(\lambda) = D_0 + \frac{D_1}{1 + \lambda} + \frac{D_2}{1 - \lambda} \equiv 0 , \tag{5.24}$$

with $D_0, D_1, D_2$ independent of $\lambda$. But for $\lambda = \lambda_0 = (c-b)/(c+b)$, (5.21) reduces to (5.20) and so $D((c-b)/(c+b)) = 0$. Similarly $D((b-c)/(b+c)) = 0$, by taking transposes. On the other hand, for $\lambda = \infty$, (5.21) reduces to showing

$$0 = \Delta[\Gamma, \Delta] - s\delta[\beta, x \wedge y] .$$

But this follows from (5.19) and (5.23), as before. Thus (5.21) holds for all $\lambda$.

From (5.5), we have

$$\begin{aligned}\frac{\det(A(\lambda)-z)}{\det(\Lambda - z)} &= 1 - \Big(\frac{2\delta}{1-\lambda^2} Q_z(x,y) + \frac{\delta^2}{1-\lambda^2}\big(Q_z(x)Q_z(y) - Q_z^2(x,y)\big)\Big) \\ &= 1 + \frac{\Delta}{s^2}\frac{1}{1-\lambda^2}\Phi_z(x,y) = 1 + \frac{\Delta}{s^2}\frac{1}{1-\lambda^2}\Big(1 - \frac{\ell(z)}{a(z)}\Big) ,\end{aligned}$$

and the curve

$$\det(A(\lambda) - z) = 0$$

is equivalent to

$$\lambda^2 = \frac{(a(s^2+\Delta) - \Delta\ell)a}{(sa)^2}$$

But $s^2 + \Delta = r^2$, and setting $\omega = \lambda s a$, we again obtain (5.12),

$$\omega^2 = (a\,r^2 - \Delta\ell)a .$$

Note that in all cases $A(\lambda)$ is self-adjoint on $iR$. This means that $e^{tF'(A(\lambda))}$ is positive definite for all real valued functions $F'(\cdot)$ on spec $A(\lambda)$, and hence the factorization

$$\begin{aligned} e^{t\,F'(A(\lambda))} &= g_+(t,\lambda)\,g_-(t,\lambda) \\ g_+(t,\infty) &= g_-(t,\infty) \end{aligned} \tag{5.25}$$

of type (1.38) will always exist and give rise to deformations of the type

$$\frac{dA(t,\lambda)}{dt} = \big[(\pi_- F'(A(t,\cdot)))(\lambda)\ ,\ A(t,\lambda)\big] \tag{5.26}$$

as before. The matrix $A(\lambda)$, however, may not be positive definite: this means in particular that the Cholesky flow ($F'(\cdot) = \log(\cdot)$) may be singular, depending on the values of $x$ and $y$.

To complete the proof of Theorem 5.14, we must show that the flows (5.8) generated by Hamiltonians of type (5.7), also arise from a loop group factorization as in (5.25), (5.26) above. From (5.7), we see that

$$H = \frac{1}{4\pi i}\int_{|z|=R} \det(L(x,y)-z)\frac{f(z)}{\det(A-z)}\,dz + \text{const.}$$

A moment's thought shows that, from the loop group point of view, the critical fact about $H$ is that it is invariant on replacing $L$ by $gLg^{-1}$ for any invertible matrix $g$. From this it is clear that, up to some tedious combinatorics, such Hamiltonians $H$ can be expressed in terms of loop group Hamiltonians of the form

$$\int_{i\mathbb{R}} \operatorname{tr}\, F(A(\lambda))\,\frac{2\,đ\lambda}{1-\lambda^2}\,,$$

and the desired result would follow by calculations similar to those in Sections 2 and 4. We will illustrate this procedure in a particular example, and leave the general case to the reader.

We will show that the loop group Hamiltonian

$$H(A) = \frac{1}{2}\int_{-i\infty}^{i\infty} \operatorname{tr}\, A^2(\lambda)\,\frac{2\,đ\lambda}{1-\lambda^2} \tag{5.27}$$

generates the flow (5.20) in the case

$$\beta = \frac{1}{b+c}\Lambda\,, \quad \text{i.e.} \quad f(\alpha_i) = \alpha_i/(b+c)\,. \tag{5.28}$$

If $b = 1+\sqrt{3}$, $c = 1-\sqrt{3}$, this corresponds to the periodic Toda lattice (see [M], also [DLTr]). Now as in Section 2, $H(A)$ generates the flow

$$\begin{aligned}\frac{dA(\lambda)}{dt} &= [\pi_- A(\lambda)\,,\, A(\lambda)]\\ &= \int [A(\lambda')\,,\, \frac{A(\lambda)-A(\lambda')}{\lambda'-\lambda}]\,đ\lambda'\\ &= \frac{\delta}{1+\lambda}\int [A(\lambda'),\, x\otimes y]\,\frac{đ\lambda'}{1+\lambda'} - \frac{\delta}{1-\lambda}\int [A(\lambda'),\, y\otimes x]\frac{đ\lambda'}{1-\lambda'}\,,\end{aligned}$$

which is equivalent to

$$\frac{d}{dt} x \otimes y = \int \left[A(\lambda') \, , x \otimes y\right] \frac{đ\lambda'}{1+\lambda'} \, , \tag{5.29}$$

by taking adjoints and again using $A^T(\lambda') = A(-\lambda')$. Substituting for $A(\lambda')$, we find eventually,

$$\frac{d}{dt} x \otimes y = \frac{1}{2}\left[\Lambda \, , x \otimes y\right] + \frac{\delta}{2}\left[y \otimes x \, , x \otimes y\right] . \tag{5.30}$$

On the other hand, letting $\lambda \to -1$ in (5.21), we obtain

$$\frac{d}{dt} x \otimes y = \left[s\beta + \Delta\Gamma, \, x \otimes y\right] . \tag{5.31}$$

But from (5.11) and (5.28),

$$\beta = \frac{1}{2s}\Lambda \, , \qquad \Gamma = \frac{1}{2s} x \wedge y \, ,$$

and (5.31) reduces to (5.30). This completes the proof of the theorem. □

**Remark 5.32.** If we represent $L$ in case (i) by

$$L = \Lambda + a \, x \otimes x + y \otimes y + r \, x \wedge y \, , \quad a \neq 0 \, ,$$

then we must consider loops of the form

$$A(\lambda) = \Lambda + \frac{a+r^2}{1-\lambda^2} x \otimes x - \frac{a+r^2}{\sqrt{-a}} \frac{\lambda}{1-\lambda^2} x \wedge y + \frac{a+r^2}{a} \frac{1}{1-\lambda^2} y \otimes y \tag{5.33}$$

and (5.21) is replaced by

$$\frac{dA(\lambda)}{dt} = \left[\sqrt{-a}\lambda\beta + \Delta\Gamma \, , A(\lambda)\right] . \tag{5.34}$$

**Remark 5.35.** On the one hand, in [M], Moser shows that the matrix $y \otimes y + x \wedge y$ is connected with geodesic flow on the ellipse $(x, \Lambda x) + 1 = 0$ (equivalently, free motion constrained to the ellipse). On the other hand, in [MV], the authors show that the matrix $y \otimes y + \lambda \, x \wedge y - \lambda^2 C^2$ is connected with the billiard ball problem in $(x, C^{-2}x) \leq 1$ ($-C^{-2}$ is just $\Lambda$ restricted to the appropriate hyperplane). But in [M], Moser also shows that the matrix $\Lambda + a\,x \otimes x + y \otimes y + r\,x \wedge y$ is connected with the motion $\ddot{x} = -ax$ in a central force field, constrained to $(x, \Lambda x) + 1 = 0$. One might then imagine that a Moser-Veselov type factorization for $(1 - \lambda^2)\,A(\lambda)$ in (5.33), would describe the billiard ball problem on an ellipsoidal billiard table where the motion *between* bounces is given by $\ddot{x} = -ax$. Unfortunately this is not the case, a fact which we find to be extremely counterintuitive.

**Remark 5.36.** A more algebraic loop group approach to the results in [M] is presented in [AHP] and also in [A].

**Remark 5.37.** The periodic Toda lattice can also be interpreted in terms of a different loop group in the following way.

Let $\underline{g}^{\#}$ be the loop Lie algebra over the circle $|z| = 1$.

$$\underline{g}^{\#} = \{A(z) = \sum_{-\infty}^{\infty} A_j z^j : A_j \in g\ell(n, R),\ A_j \to 0 \text{ exponentially as} |j| \to \infty\} ,$$

with pointwise commutator

$$[A, B](z) = [A(z), B(z)] .$$

Set

$$(P_+A)(z) = \sum_{j>0} A_j z^j , \qquad (P_-A)(z) = \sum_{j<0} A_j z^j ,$$

and define

$$\underline{k}^{\#} = \{A \in \underline{g}^{\#} : A^*(z) + A(z) = 0\}$$

$$\underline{\ell}^{\#} = \{A + \underline{g}^{\#} : (P_+A)(z) = 0,\ A_0 = 0\} .$$

Clearly we have the direct decomposition

$$\underline{g}^{\#} = \underline{\ell}^{\#} + \underline{k}^{\#}$$

with associated projections

$$\pi_{\underline{\ell}^{\#}} A(z) = \pi_\ell A_0 + (P_-A)(z) + (P_+A)(z)^*$$

$$\pi_{\underline{k}^{\#}} A(z) = \pi_k A_0 + (P_+A)(z) - (P_+A)(z)^* ,$$

where $\pi_k X = X_+ - X_+^T$, $\pi_\ell X = (1 - \pi_k)X$ for any matrix $X$.

Again

$$R^{\#} = \pi_{\underline{\ell}^{\#}} - \pi_{\underline{k}^{\#}}$$

solves the modified Yang-Baxter equation, giving rise to the Lie algebra $\{\tilde{\underline{g}}_{R^{\#}}, [\cdot,\cdot]_{R^{\#}}\}$ with associated group $\widetilde{G}_{R^{\#}}$.

The group $\widetilde{G}_{R^{\#}}$ has particular $2N-2$ dimensional coadjoint orbits $O_A \subset \tilde{\underline{g}}^*_{R^{\#}}$ of the form

$$A(z) = A_1^T z^{-1} + A_0 + A_1 z$$

where

$$A_0 = \begin{pmatrix} a_1 & b_1 & & & O \\ b_1 & & & & \\ & \ddots & \ddots & \ddots & \\ & & & & b_{N-1} \\ O & & & b_{N-1} & a_N \end{pmatrix}, \qquad A_1 = \begin{pmatrix} 0 & O \\ b_n & 0 \end{pmatrix}$$

and

$$\sum_{i=1}^{N} a_i , \qquad \prod_{i=1}^{N} b_i$$

are fixed. Moreover the Flaschka map

$$\begin{aligned}(R^{2N}, \sum_{i=1}^{N} dx_i \wedge dy_i) \ni (x,y) &\to \left(a_k = -y_k/2,\ b_k = \frac{1}{2}e^{(1/2)(x_k - x_{k+1})}\right) \\ &\to A \in (\tilde{\underline{g}}^*_{R^{\#}}, \{\cdot,\cdot\}_{R^{\#}})\end{aligned}$$

is symplectic. Note that, as opposed to the orbit $\tilde{\underline{g}}^*_{\text{sing}}$ in Section 2, elements in $O_A$ do *not* have a polar part.

Given initial data $(a_k^0, b_k^0) \leftrightarrow A^0 = (A_1^0)^T z^{-1} + A_0^0 + A_1^0 z$, the periodic Toda lattice can be solved via the factorization problem arising from the Hamiltonian

$$H_{Toda} = \frac{1}{2} \int_{|z|=1} \text{tr } A(z)^2 \frac{dz}{z}$$

in the usual way:

$$e^{tA^0(z)} = g_{\underline{\ell}^{\#}}(t,z)\, g_{\underline{k}^{\#}}(t,z) ,$$

where

$$g_{\underline{\ell}^\#}(t,\cdot) \in e^{\pi_{\underline{\ell}^\#} \underline{g}^\#}, \quad g_{\underline{k}^\#}(t,\cdot) \in e^{\pi_{\underline{k}^\#} \underline{g}^\#},$$

and

$$\begin{aligned} A(t,z) &= \left(g_{\underline{\ell}^\#}(t,z)\right)^{-1} A^0(z)\, g_{\underline{\ell}^\#}(t,z) \\ &= A_1^T(t) z^{-1} + A_0(t) + A_1(t) z \in O_{A^0} . \end{aligned}$$

For the general problem considered in the text, however, we have not been able to discover a loop group whose coadjoint orbits are relevant to the problem at hand, and at the same time do not contain a singular part.

## Appendix. Classical $R$-matrix theory

Classical $R$-matrix theory originated in the work of Adler, Kostant and Symes (see, for example, [A] or [Sy1]), and has been developed extensively over recent years, particularly by the Russian school in Leningrad (see, for example, [RSTS], [FT]). The theory shows how to construct commuting integrals for a wide class of Hamiltonian flows, and also shows how to integrate the flows explicitly by factorization. In this Appendix we give a brief sketch of the relevant results from the theory: proofs and systematic accounts can be found, for example, in [FT] and in [STS]. In the presentation that follows we have in mind the finite dimensional case. In the infinite dimensional case, the results remain true after suitable modifications, as described, for example, in the generalized Cholesky algorithm of the text.

Let $\underline{g}$ be a Lie algebra with bracket $[\cdot,\cdot]$. A linear operator $R : \underline{g} \to \underline{g}$ is said to be a *classical R-matrix* if the bracket on $\underline{g}$ given by

$$[X, Y]_R = \frac{1}{2}([RX, Y] + [X, RY]) \tag{A.1}$$

is a Lie-bracket, i.e. satisfies the Jacobi identity. We use the symbol $\underline{g}_R$ to denote the algebra $\underline{g}$ equipped with the bracket $[\cdot,\cdot]_R$. Along with the two Lie-brackets on $\underline{g}$ we shall consider the corresponding Lie-Poisson brackets on $C^\infty(\underline{g}^*)$,

$$\{F_1, F_2\}(\alpha) = \alpha([X_1, X_2]) \tag{A.2}$$

and

$$\{F_1, F_2\}_R(\alpha) = \alpha([X_1, X_2]_R)\ , \qquad \alpha \in \underline{g}^*\ ,\ X_i = dF_i(\alpha) \in \underline{g}\ . \tag{A.3}$$

The symbols $ad^*$ and $ad^*_R$ stand, respectively, for the coadjoint representations of the algebras $\underline{g}$ and $\underline{g}_R$. It follows readily that

$$(ad^*_R X)(\alpha) = \frac{1}{2}((ad^* RX)(\alpha) + R^*(ad^* X(\alpha)))\ . \tag{A.4}$$

An interesting and useful sufficient condition for $[\cdot,\cdot]_R$ to be a Lie-bracket is the so-called modified Yang-Baxter equation

$$[RX, RY] - 2R([X,Y]_R) = -[X,Y]\ , \quad X, Y \in \underline{g} \tag{A.5}$$

The connection between the modified Yang-Baxter equation and factorization theory is made clear by noting that

$$[(R \pm 1)X,\, (R \pm 1)Y] = 2(R \pm 1)([X,Y]_R) \tag{A.6}$$

is equivalent to (A.5). From this reformulation of (A.5) we see immediately that $R$ gives rise to a (not necessarily direct) decomposition of $\underline{g}$,

$$\underline{g} = \underline{g}_+ + \underline{g}_- \tag{A.7}$$

into a sum of *subalgebras*

$$\underline{g}_\pm \equiv (1 \pm R)\underline{g}\ . \tag{A.8}$$

Set

$$\pi_\pm = \frac{1}{2}(1 \pm R) \tag{A.9}$$

so that

$$I = \pi_+ + \pi_- \tag{A.10}$$

and

$$R = \pi_+ - \pi_-\ . \tag{A.11}$$

If the decomposition in (A.7) is direct, then $\pi_\pm$ are the complementary projections onto $\underline{g}_\pm$ respectively. Conversely, if $\underline{g} = \underline{g}_+ \dotplus \underline{g}_-$ is a direct decomposition of $\underline{g}$ into subalgebras $\underline{g}_+$ and $\underline{g}_-$ with associated projections $\pi_\pm$, then $R \equiv \pi_+ - \pi_-$ solves the modified Yang-Baxter equation.

Let $G$ be a Lie group with Lie algebra $\underline{g}$ and co-adjoint action $Ad^*$, and let $G_\pm$ be subgroups of $G$ corresponding to the Lie subalgebras $\underline{g}_\pm = \pi_\pm \underline{g}$. We say that a smooth function $F : \underline{g}^* \to \mathbb{C}$ is $Ad^*$-invariant if

$$F(Ad^*_g \alpha) = F(\alpha) \tag{A.12}$$

for all $\alpha \in \underline{g}^*$ and $g \in G$. By differentiation, this implies

$$ad^*(dF(\alpha))(\alpha) = 0 \quad \text{for all} \quad \alpha \in \underline{g}^* . \tag{A.13}$$

A smooth function $F : \underline{g}^* \to \mathbb{C}$ generates the flow

$$\dot{\alpha} = \{F, \alpha\}_R = \frac{1}{2}(ad^*(RdF(\alpha))(\alpha) + R^* ad^*(dF(\alpha))(\alpha) , \tag{A.14}$$

and when $F$ is $Ad^*$-invariant,

$$\dot{\alpha} = \frac{1}{2} ad^*(R\, dF(\alpha))(\alpha) = -ad^*(\pi_- dF(\alpha))(\alpha) \tag{A.15}$$

where we have used (A.13) twice and (A.9) once. In the case that $(\cdot, \cdot)$ is a nondegenerate ad-invariant pairing on $\underline{g}$,

$$(x, ad\, Y(Z)) + (ad\, Y(X), Z) = 0 , \tag{A.16}$$

and $\underline{g}^*$ is identified with $\underline{g}$ through the pairing, then (A.15) takes on the Lax-pair form

$$\dot{\alpha} = [\pi_- dF(\alpha) , \alpha] . \tag{A.17}$$

The following are the basic results.

**Theorem A.18.1.** *$Ad^*$-invariant functions Poisson commute on $(\underline{g}^*_R, \{\cdot, \cdot\}_R)$.* □

**Theorem A.18.2.** *Suppose that $F$ is $Ad^*$-invariant and that the decomposition in* (A.7) *is direct. Then for some $T \le \infty$, there exists a unique decomposition*

$$e^{t\, dF(\alpha_0)} = g_+(t) g_-(t) \tag{A.19}$$

*for* $0 \le t < T$, *where* $g_\pm(t) \in \exp \underline{g}_\pm$, *and*

$$\alpha(t) = Ad^*_{g_+(t)}\alpha_0 = Ad^*_{g_-(t)^{-1}}\alpha_0 \tag{A.20}$$

*is the solution of* (A.17) *for* $0 \le t < T$ *with* $\alpha(0) = \alpha_0$.

**Proof:** Given (A.19), we show that (A.20) solves (A.17). By differentiation, (A.20) implies

$$\frac{d}{dt}\alpha(t) = -ad^*(g_-'(t)\,g_-(t)^{-1})(\alpha(t)) \tag{A.21}$$

But from (A.19),

$$Ad_{g_+(t)^{-1}}dF(\alpha) = g_+(t)^{-1}\,g_+'(t) + g_-'(t)\,g_-(t)^{-1} \ . \tag{A.22}$$

On the other hand, differentiation of (A.12) with respect to $\alpha$, yields

$$dF(Ad_g^*\alpha) = Ad_{g^{-1}}(dF(\alpha)) \ , \tag{A.23}$$

and inserting this relation in (A.22), we obtain

$$dF(\alpha(t)) = g_+(t)^{-1}\,g_+'(t) + g_-'(t)\,g_-^{-1}(t) \ . \tag{A.24}$$

But the decomposition in (A.7) is direct and so

$$\pi_- dF(\alpha(t)) = g_-'(t)\,g_-(t)^{-1} \ . \tag{A.25}$$

Substitution in (A.21) now gives the desired result. □

In the case that the decomposition in (A.7) is not direct, the proof of formula (A.25) fails and the approach must be modified. Following [STS], we note first that

$$K_\pm = \mathrm{Ker}(R \mp 1) \tag{A.26}$$

are ideals in $\underline{g}_\pm$. Furthermore the map $\theta : \underline{g}_+/K_+ \to \underline{g}_-/K_-$ taking

$$(R+1)x + K_+ \mapsto (R-1)x + K_- \tag{A.27}$$

is a well defined Lie algebra isomorphism, which lifts to a Lie group homomorphism $\Theta$ from $e^{\underline{g}_+}/e^{K_+} \to e^{\underline{g}_-}/e^{K_-}$.

**Theorem A.18.3.** *Theorem* A.18.2 *remains true in the case that the decomposition* (A.7) *is not direct, provided that* $g_+(t)$ *and* $g_-(t)$ *satisfy*

$$\Theta([g_+(t)]) = [g_-(t)^{-1}] , \tag{A.28}$$

*where* $[\cdot]$ *denotes the equivalence classes in* $e^{\underline{g}_\pm}/e^{K_\pm}$.

**Proof:** It is enough to verify (A.25). But by the homomorphism property and (A.28),

$$\Theta[g_+(t)^{-1}\, g_+(t+s)] = [g_-(t)\, g_-(t+s)^{-1}]$$

for all $t, s \geq 0$, and so

$$\theta([g_+(t)^{-1}\, g_+'(t)]) = [-g_-'(t)\, g_-^{-1}(t)]$$

by differntiation. Hence, by (A.27), there exist $x \in \underline{g}$ and $k_\pm \in K_\pm$, such that

$$\begin{aligned} g_+(t)^{-1}\, g_+'(t) &= (R+1)x + k_+ \\ -g_-'(t)\, g_-^{-1}(t) &= (R-1)x + k_- \, . \end{aligned} \tag{A.29}$$

Substituting in (A.24), we obtain

$$dF(\alpha(t)) = 2x + k_+ - k_- ,$$

which yields

$$\begin{aligned} \pi_- dF(\alpha(t)) &= 2(\frac{1-R}{2})x + \frac{(1-R)}{2}k_+ - \frac{1-R}{2}k_- \\ &= (1-R)x - k_- \\ &= g_-'(t)\, g_-^{-1}(t) \, . \end{aligned}$$

as desired. □

In the classical Cholesky algorithm (see [DLT]), $\underline{g} = g\ell(N, \boldsymbol{R})$ and

$$R(M) = M_+ - M_- , \qquad M \in \underline{g} , \tag{A.30}$$

where $M_\pm$ denote the strict upper/lower triangular parts of $M$ respectively. One checks that $\underline{g}_\pm$ are the upper/lower triangular matrices and $K_\pm$ are the strictly upper/strictly lower triangular matrices: thus $\underline{g}_\pm / K_\pm \cong$ diagonal matrices and $e^{\underline{g}_\pm} / e^{K_\pm} \cong$ positive, invertible diagonal matrices. Furthermore, $\theta$ takes the diagonal matrix $\delta$ to $-\delta$ and the condition (A.28) reduces in this case to the requirement that

$$\operatorname{diag} g_+(t) = \operatorname{diag} g_-(t) , \tag{A.31}$$

which is precisely the condition appearing in [DLT].

**Bibliography**


[AHP] M. R. Adams, J. Harnad and E. Previato, Isospectral Hamiltonian flows in finite and infinite dimensions I. Generalized Moser systems and moment maps into loop algebras, Comm. Math. Phys., 117, 1988, 451-500.

[A] M. Adler, On a trace functional for formal pseudo-differential operators and the symplectic structure of the Korteweg-deVries equations, Inv. Math., 50, 1979, 219-248.

[BDT] R. Beals, P. Deift and C. Tomei, Direct and inverse scattering on the line, Math. Surveys and Monographs, No. 28, AMS, Providence, RI, 1988.

[B] G. D. Birkhoff, Dynamical Systems, Colloquium Publications, Vol. 9, A.M.S., Providence, RI, 1927 (revised edition, 1966).

[BGM] A. Bunse-Gerstner and V. Mehrmann, A symplectic QR like algorithm for the solution of the real algebraic Riccati equation, IEEE Trans. Auto. Contr., Vol. AC-31, No. 12, 1986, 1104-1113.

[Chu] M. T. Chu, On the continuous realization of iterative processes, SIAM Review, 30 , 1988, 375-387.

[DL] P. Deift and L. C. Li, Generalized affine Lie algebras and the solution of a class of flows associated with the QR eigenvalue algorithm, Comm. Pure Appl. Math., 42, 1989, 963-991.

[DLNT] P. Deift, L. C. Li, T. Nanda and C. Tomei, The Toda flow on a generic orbit is integrable, Comm. Pure Appl. Math., 39, 1986, 183-232.

[DLT] P. Deift, L. C. Li and C. Tomei, Matrix factorizations and integrable systems, Comm. Pure Appl. Math., 42, 1989, 443-521.

[DLTr] P. Deift, F. Lund and E. Trubowitz, Nonlinear wave equations and constrained harmonic motion, Comm. Math. Phys., 74, 1980, 141-188.

[DT] P. Deift and E. Trubowitz, Inverse scattering on the line, Comm. Pure Appl. Math., 32, 1979, 121-251.

[DTT] P. Deift, C. Tomei and E. Trubowitz, Inverse scattering and the Boussinesq Equation, Comm. Pure Appl. Math., 35, 1982, 567-628.

[FT] L. D. Faddeev and L. A. Takhtajan, Hamiltonian methods in the theory of solitons, Springer-Verlag, Berlin, 1987.

[GK] I. C. Goh'berg and M. G. Krein, Systems of integral equations on a half-line with kernels depending on the difference of the arguments, AMS Translations, No. 2, Vol. 14, 1960, 217-287.

[GM] V. Guillemin and R. Melrose, An inverse spectral result for elliptic regions in $\boldsymbol{R}^2$, Adv. in Math., 32, 1979, 128-148.

[K] T. Kato, Perturbation theory for linear operators, Springer-Verlag New York Inc., New York, 1966.

[Man] S. V. Manakov, Remarks on the integration of the Euler equations of an $n$-dimensional rigid body, Funct. Anal. Appl., 10, 1976, No. 4 93-94 (Russian).

[M] J. Moser, Geometry of quadrics and spectral theory, The Chern Symposium, Berkeley, June 1979, Springer-Verlag, New York, Heidelberg, Berlin, 1980, 147-188.

[MV] J. Moser and A. P. Veselov, Discrete versions of some classical integrable systems and factorization of matrix polynomials, Preprint, ETH, Zürich, 1989.

[RSTS] A. G. Reyman and M. A. Semenov-Tian-Shanskii, Reduction of Hamiltonian systems, affine Lie algebras and Lax equations II, Invent. Math., 63, 1981, 423-432.

[S] R. Schilling, Genealizations of the Neumann system. A curve-theoretical approach — Part I, Comm. Pure Appl. Math., 40, 1987, 455-522.

[STS] M. A. Semenov-Tian-Shanskii, What is a classical $R$-matrix? Func. Anal. Prilož., 17(4), 1983, 17-33 (Russian); English translation, Funct. Anal. Appl., 17, 1984, 259-272.

[SH] M. A Singer and S. J. Hammarling, The algebraic Riccati equation: a review of some available results, NPL Rep. DITC 23/83, 1983.

[Sy1] W. W. Symes, Hamiltonian group actions and integrable systems, Physica 1D, 1980,

339-374.

[Sy2] W. W. Symes, The QR algorithm and scattering for the finite nonperiodic Toda lattice, Physica 4D, 1982, 275-280.

[Th] A. Thimm, Integrable geodesic flows on homogeneous spaces, Ergod. Th. and Dynam. Sys., 1, 1981, 495-517.

[V] A. P. Veselov, Integrable systems with discrete time and difference operators, Funct. Anal. and Appl., 22, 1988, No. 2, 1-13 (Russian).

[Wa] D. Watkins, Isospectral flows, SIAM Rev., 26, 1984, 379-391.

[Wi] J. Wilkinson, The algebraic eigenvalue problem, Oxford Univ. Press, London, 1965.

P. A. Deift, Courant Institute, New York,

L. C. Li, Pennsylvania State University, University Park

C. Tomei, PUC, Rio de Janeiro

## Editorial Information

To be published in the *Memoirs*, a paper must be correct, new, nontrivial, and significant. Further, it must be well written and of interest to a substantial number of mathematicians. Piecemeal results, such as an inconclusive step toward an unproved major theorem or a minor variation on a known result, are in general not acceptable for publication. *Transactions* Editors shall solicit and encourage publication of worthy papers. Papers appearing in *Memoirs* are generally longer than those appearing in *Transactions* with which it shares an editorial committee.

As of September 1, 1992, the backlog for this journal was approximately 9 volumes. This estimate is the result of dividing the number of manuscripts for this journal in the Providence office that have not yet gone to the printer on the above date by the average number of monographs per volume over the previous twelve months. (There are 6 volumes per year, each containing about 3 or 4 numbers.)

A Copyright Transfer Agreement is required before a paper will be published in this journal. By submitting a paper to this journal, authors certify that the manuscript has not been submitted to nor is it under consideration for publication by another journal, conference proceedings, or similar publication.

## Information for Authors

*Memoirs* are printed by photo-offset from camera copy fully prepared by the author. This means that the finished book will look exactly like the copy submitted.

The paper must contain a *descriptive title* and an *abstract* that summarizes the article in language suitable for workers in the general field (algebra, analysis, etc.). The *descriptive title* should be short, but informative; useless or vague phrases such as "some remarks about" or "concerning" should be avoided. The *abstract* should be at least one complete sentence, and at most 300 words. Included with the footnotes to the paper, there should be the 1991 *Mathematics Subject Classification* representing the primary and secondary subjects of the article. This may be followed by a list of *key words and phrases* describing the subject matter of the article and taken from it. A list of the numbers may be found in the annual index of *Mathematical Reviews*, published with the December issue starting in 1990, as well as from the electronic service e-MATH [**telnet e-MATH.ams.org** (or **telnet 130.44.1.100**). Login and password are **e-math**]. For journal abbreviations used in bibliographies, see the list of serials in the latest *Mathematical Reviews* annual index. When the manuscript is submitted, authors should supply the editor with electronic addresses if available. These will be printed after the postal address at the end of each article.

**Electronically-prepared manuscripts.** The AMS encourages submission of electronically-prepared manuscripts in $\mathcal{A}_{\mathcal{M}}\mathcal{S}$-TEX or $\mathcal{A}_{\mathcal{M}}\mathcal{S}$-LATEX. To this end, the Society has prepared "preprint" style files, specifically the amsppt style of $\mathcal{A}_{\mathcal{M}}\mathcal{S}$-TEX and the amsart style of $\mathcal{A}_{\mathcal{M}}\mathcal{S}$-LATEX, which will simplify the work of authors and of the production staff. Those authors who make use of these style files from the beginning of the writing process will further reduce their own effort.

*Guidelines for Preparing Electronic Manuscripts* provide additional assistance and are available for use with either AMS-TEX or AMS-LATEX. Authors with FTP access may obtain these *Guidelines* from the Society's Internet node e-MATH.ams.org (130.44.1.100). For those without FTP access they can be obtained free of charge from the e-mail address guide-elec@math.ams.org (Internet) or from the Publications Department, P. O. Box 6248, Providence, RI 02940-6248. When requesting *Guidelines* please specify which version you want.

Electronic manuscripts should be sent to the Providence office only after the paper has been accepted for publication. Please send electronically prepared manuscript files via e-mail to pub-submit@math.ams.org (Internet) or on diskettes to the Publications Department address listed above. When submitting electronic manuscripts please be sure to include a message indicating in which publication the paper has been accepted.

For papers not prepared electronically, model paper may be obtained free of charge from the Editorial Department at the address below.

Two copies of the paper should be sent directly to the appropriate Editor and the author should keep one copy. At that time authors should indicate if the paper has been prepared using AMS-TEX or AMS-LATEX. The *Guide for Authors of Memoirs* gives detailed information on preparing papers for *Memoirs* and may be obtained free of charge from AMS, Editorial Department, P. O. Box 6248, Providence, RI 02940-6248. The *Manual for Authors of Mathematical Papers* should be consulted for symbols and style conventions. The *Manual* may be obtained free of charge from the e-mail address cust-serv@math.ams.org or from the Customer Services Department, at the address above.

Any inquiries concerning a paper that has been accepted for publication should be sent directly to the Editorial Department, American Mathematical Society, P. O. Box 6248, Providence, RI 02940-6248.

W9-CXE-927

# THE DOG BOOK

*By*

DIANA THORNE

*and*

ALBERT PAYSON TERHUNE

THE SAALFIELD PUBLISHING COMPANY

AKRON, OHIO NEW YORK

MADE IN U. S. A.

# Contents

# Color Illustrations

# THE DOG BOOK

---

## The Collie

THOUSANDS of years ago a shaggy little coal-black dog trotted clear across Europe from the Near East. He was not half the size of the collies you see nowadays; and probably he was not much to look at. But he was useful. That was why he crossed Europe.

He and other little black dogs like him were herding the enormous droves of cattle and sheep which formed the wealth of a dark-faced crowd of Orientals. These dark-faced men were invaders. They plundered and fought as they went. At last some of them chose the northern part of the British Isles as their new home. That is how the collie found his way to Great Britain. And there his descendants have been herding sheep and cattle ever since. Because of their jet-black color they became known as "coalies"; and many Scotchmen still call them that, instead of their present name collie.

In Scotland and in England—everywhere that sheep were kept—the collie was kept, too. For no other dog understood herding as did he. He was wise, and he was fearless and tireless, and he was loyal to the death. Soon these fine traits made him the housemate of farmers and shepherds, who found him the grandest four-legged pal in the world. He had a place of honor on his master's hearth; and he even lay quietly in the aisle beside his master's pew in church. Often, in Scotland, I have seen big collies lying gravely at their owner's feet throughout long church services. They behaved better than did some of the human members of the congregation.

As time went on, the collie was not always black. Oftener he was yellow or brown or gray or even white. Queen Victoria made him fashionable by having a handsome collie as her daily companion and guard, preferring him to any other kind of dog. Presently thousands of collies were kept as pets, not only in England but in America, too.

This rush of popularity came near spoiling a wonderful dog. Collies are meant to work or to be the outdoor pals of outdoor people. Fashion said a collie must have a longer nose and a thinner head and smaller eyes than nature had meant him to have. So they proceeded to "breed his brains out through his nose." In other words, to make him almost a freak. Also he was often kept cooped up in a city apartment or in a small back yard, where he had no chance for the long sweeping runs that are so needful to a collie. It was like putting a hawk in a canary cage. The big city is no place for big dogs.

Then some of us started to bring the collie back to what he used to be — vigorous and clever and strong. He stopped his slide downward to freakishness. He was saved. His queerly wise brain also got a chance to develop. I think a collie's brain and nature are not like those of any other dog. There is an elfin gaiety and a sense of fun, mingled with staunch loyalty, in a collie, that is not found elsewhere.

I had a huge "merle" collie, named Gray Dawn. We taught him to go to the gate every morning when the newsdealer left the papers there, and to bring them to the house. He learned this quickly; and when we praised him for it, he was so proud that he decided to improve on the simple trick. Next morning there were more than twenty newspapers on our front porch. Gray Dawn had gone around the neighborhood and had gathered up all the papers left by the newsdealer at other people's doors, and had brought them home to us. I doubt if any dog but a collie would have done that.

Once when I was injured by a motor-car, Gray Dawn would not leave my bedside, day or night. For an entire week he refused to eat. When first the doctor laid a hand on my injured arm, Gray Dawn leaped murderously at the throat of the man who he thought was hurting his master.

Then there was our great Sunnybank Lad, who had almost human reasoning powers. Lad was the sire of our fiery little red-gold collie, Wolf. When Wolf was a puppy, he made old Laddie's life a burden by following him around and watching where Lad buried bones. Then as soon as Lad's back was turned, Wolf would dig up the bone and carry it away to gnaw.

Before long, Laddie thought out a way to stop this. He dug an unusually deep hole in the ground and put a large and juicy bone at the bottom of it. He shoved an inch or more of dirt over this bone. Then above it he placed a worthless and meatless and weather-stained old bone he had found, a bone no dog could possibly be interested in. After that, he shoved the rest of the dirt back into the cavity and went away.

Immediately young Wolf came over to rob the hiding place. He dug down until he came upon the worthless, old dry bone. He gave one contemptuous sniff at it and walked off in disgust, never guessing there was a new and appetizing bone buried just an inch or so beneath the decoy bone. If that was not a case of clever reasoning on Lad's part, I don't know what reasoning is.

By the way, do you know why dogs bury bones? Except for squirrels and chipmunks, I think they are the only animals to bury food for future use. In prehistoric days, wild dogs had to depend on other and weaker creatures for a living. They would catch and kill a goat, or a deer, or some other animal, and make a meal of it. Often there was much more than they could eat at one time. They dared not leave the remainder of the carcass lying there until they should be hungry again. Some other dog wandering in search of food might scent it out and devour it. So they buried the bones where they could come back and dig them up when they wished. The scent would not penetrate the earth, and the food would be safe from marauders. This habit has been passed along through the centuries to modern dogs, especially to collies.

A dog always digs the holes for burying bones with his forepaws. But he never uses them to fill the earth in again. He always does this with his nose. I wonder why.

In the World War countless dogs were used on the battle-front for carrying messages and for many other purposes. None of these four-footed messengers were better than the collies. Here is an account taken from the records of the British War Office.

A detachment of troops was cut off from the main body of the British army. The enemy were about to attack them in overwhelming numbers. Unless word could be gotten to Headquarters in time for reinforcements to come to their aid, they would be wiped out. The telephone had been destroyed. The only way to get help was to send a messenger. But a long and high hill lay between the detachment and Headquarters. There were no trees or even bushes on this hill. Any man running up its bare slope would be picked off immediately by enemy sharpshooters.

The colonel wrote out a message and tied it to the collar of a collie that had been trained to carry dispatches. Then the dog was lifted over the top of the trench and told where to go. The moment he appeared on the hillside, the enemy sharpshooters began to fire at him. But a fast-running collie is hard to hit. Through a hailstorm of bullets the dog galloped at top speed. A bullet raked his side, inflicting a painful flesh-wound, but he did not so much as swerve in his stride. Another bullet cut his cheek. The sharpshooters were getting the range at last.

As he neared the crest of the hill, a bullet tore clean through his body. He lurched to his feet, and ran on. As he reached the crest, a bullet broke his spine. He slumped to the ground, apparently dead. But presently he had upreared his forelegs, and began to crawl onward, dragging his hind legs behind him. He disappeared over the top of the hill.

Half an hour later the collie crept painfully into Headquarters. The gallant dog's work was done. Now he could afford to die. And he died while the message was being taken from his collar — the message which saved hundreds of helpless men from death or surrender.

Did you ever see a litter of collie puppies? When they are born, they weigh less than a pound, and they look like blind rats. By the time they are six weeks old, they look like Teddy Bears, and are full of mischief and friendliness. To me, there are few sights half as pretty and amusing as a yard full of baby collies at play.

I have been raising collies for more than thirty years. In that time, I don't know how many hundred people have asked me: "Aren't collies treacherous?" And always I reply: "If they were, I wouldn't have them around me." I don't know where that idea of collies being treacherous originated. But I can tell you from long experience that they are *not*. I must have owned nearly a thousand collies, in my time, and out of all that number I have known only one at Sunnybank to which the term could possibly apply. I'll tell you about him, and leave you to judge whether his actions were treacherous or whether they were only part of the elfin mischief which is a collie's heritage.

His name was Boze. He was friendly and gay; and was the chum of my three little nephews. In all the years we had him, he tried to harm only one person. That was a peddler who used to come to Sunnybank now and then to sell tinware. This man had lost his right leg, and he had a wooden leg in its place.

Whenever the peddler came down the driveway toward the house, Boze used to frisk out to greet him. The collie would run up to him, wagging his tail; then dropping behind the man, would nip him sharply in the leg. But it was always in the wooden leg; never in the other. I think the wise dog knew somehow that the right leg was a fake, and that his sense of mischief made him bite it. He gave no more than that one exploratory bite; and he never bit anyone else. Was that treachery or was it fun?

# The Police Dog

HE IS known to most of the public as the police dog. And probably he will be known thus to the end of the chapter. But that is not his real name. He is the German shepherd dog. That is what professional fanciers call him, and that is his correct name. But in this book, suppose we call him by the title you are more accustomed to. Shall we?

Nearly twenty years ago, I was strolling through the Westminster Kennel Club's annual dog show at Madison Square Garden. In those times it was a four-day show, not a more merciful three-day show as now. It was on the morning of the show's first day that I happened to wander along the section where the police dogs were benched. There was not such a throng of police dogs at the show as nowadays. But there was a goodly number, and they were benched at one end of the building.

I stopped to glance at a splendid grizzled giant of a dog at one of these benches. His owner was grooming him, and fell into talk with me. He was loud in his praises of the dog's intelligence, and suggested a test to show the animal's perfect obedience.

Up to this time the great dog had paid no attention at all to me. Now, the man made certain his chain was strongly fastened. Then he said: "Pretend to strike me."

I did so, hitting right dramatically at his face, and checking my blow a bare half-inch from his nose.

Instantly the police dog was straining at his stout chain, roaring in fury and doing his best to get at me and kill me. I was thankful the chain was so strong.

"Very interesting," I commented. "But nothing unusual. Almost any of my own dogs will fly at a stranger or at anyone else who pretends to hit me. It is a common trait with loyal dogs of any breed. I don't doubt there are a hundred dogs at this show today who would go for me if they thought I was attacking their masters."

"That isn't the point," said the man. "That is only the first move in the test. You have seen that my dog hates you because he thinks you tried to hurt me; and you can see he is mad to get at you and tear you to pieces. Now then." He quieted the raging dog and spoke to him rapidly in German, a language I do not understand. Then he turned to me again, saying, "Will you go up to him now and pat him on the head? I will guarantee he won't bite you. He has had his orders from me."

With no eagerness at all, I stepped across to the dog, and stroked his classic head. I could feel him quiver like a leaf under my caressing hand. But he stood stock-still and made no move to resent the handling or to harm me in any way. That was splendid discipline, and I praised it highly.

That was on the morning of the first day of the show. I did not happen to visit that corner of the building again until the evening of the fourth day. As I passed the bench where the big police dog lay asleep, his nostrils quivered. His eyes opened. He stared at me blankly for an instant. Then I saw recognition and hatred dawn in his glance. He leaped up and in wild fury hurled himself at me to the length of his chain. Yes, he remembered me after four whole days — days in which many thousand people had passed and repassed in front of his bench. In that time he had encountered millions of new sights and smells and sounds. Through them all, he had treasured the angry memory of the man who had struck his owner.

I have told this story at some length because I think it gives a better idea of the wonderful brain-power of a police dog than any other instance I have known. There is a deep wisdom, a wealth of almost uncanny mentality, about the best type of police dog that is found in perhaps no other breed. And this mentality can be used, and has been used, in many ways.

First of all, it has been used in the way which has given to the German shepherd dog his nickname of police dog. Long ago, in police circles he was recognized as a possible ally of great value in tracking down and subduing criminals. Training schools were established where police dogs were taught scientifically how to guard property; not to flinch from blow or gunshot; to tackle crooks in such fashion as to trip up and subdue them, and to hold them helpless until human aid should arrive; to scale seemingly unscalable fences in pursuit of their prey; and to do patrol duty. In Germany and in other countries, the trained police dog has proved himself an invaluable adjunct to the force. No other breed has been able to do the work half as well as he. His brain, his memory, his courage, and his swift teachableness lend themselves to the man hunt and to maintaining law and order. He has earned his right to the name police dog many times over.

But of late years he has been trained to a far more humane task. All over Europe and America, schools have been formed where police dogs are trained to act as eyes to the blind. Perhaps you have read of some of these. Here is one of the ways it is managed.

A young police dog is taught the rules of guiding blind people. Then someone who is blind and wants such a dog visits the school every day for two or three weeks. During that time he spends many hours a day with the dog he is going to buy. He and the police dog get acquainted; and the buyer learns the way to handle the dog and to submit to its guidance. Usually a pair of stiff reins are fastened to the dog's harness, and the blind owner holds these reins. Thus he guides and is guided. The dog can pilot him through the thickest traffic, stopping at the curb, steering him clear of obstacles, leading him at command to any place they have once visited together; in short, serving as eyes to the unseeing owner and as guard as well.

Several times, tests of this kind have been made, where a trained police dog has piloted his blind master through the traffic jams at Broadway and Forty-second street in New York, keeping him from colliding with passers-by, warning him of the change in traffic commands, and so on. It is an inspiring sight to watch. Never does the wise dog make a mistake.

Thanks to these police dog schools, hundreds of formerly helpless blind persons can now go freely anywhere as safely as if some fellow-human were convoying them. Every year, more and more police dogs are sold to the blind. Every year the teaching of the dogs becomes more efficient; and the powers of the blind are thus increased. If he were worthless for all else, this guidance of the blind has given the police dog his right to live; has justified his existence.

At one time it was declared that the police dog was a direct descendant of the wolf. This may seem to be borne out by his rufous-gray coat and by his general aspect, but it is denied most earnestly by professional fanciers and students of the breed. They claim he is one of the very oldest pure types of dogs on earth, and bring forth much testimony to back their claims. Yes, it is said the police dog was on earth long before the beginning of regular history; away back in the ages that now are almost forgotten. Students of the breed say he looked then as he looks now; and that there was no taint of wolf blood in his veins.

For centuries he was used for herding sheep, in Germany and elsewhere; just as the collie was used in Great Britain. He was a grand herder and a savage foe to marauding wild beasts which came prowling around the flocks. His cleverness and his mighty strength made him a treasure to farmers in central Europe. Here he earned his name of German shepherd dog.

When the motion pictures began to use dogs, the wisdom and the ability to learn swiftly everything rightly taught to him made the police dog more eagerly sought than were dogs of any other breed. Several fine movie plays were written for police dogs; and these dogs made world-wide reputations as screen actors.

The first of them to win fame in pictures was Strongheart. A man named Larry Trimble bought the dog after he had been trained for police service. Patiently Trimble taught the great dog his new job as actor. I was invited to a preview of Strongheart's first picture, and I went, expecting to see him perform only a few cheap tricks. But I had been much mistaken. I saw the work of a wonderful actor, even if that actor happened to be a dog instead of a man or a woman. It was a revelation to me.

Then came other police dogs that made big reputations on the screen. There was Peter the Great, for instance; and there was Rin-Tin-Tin. Rin-Tin-Tin was born on a battlefield, during the World War. His master brought him to America and taught him to be a star actor.

Rightly trained, there is almost no limit to the police dog's powers to absorb education. I could tell you fifty stories to prove this, but the one I told at the beginning of this article is a good example. And, besides, it shows what absolute obedience the police dog gives to the man whom he has fully accepted as his master. These dogs have the self-discipline of a perfect soldier.

During the World War, soldiers from our country had many

chances to study police dogs. They came home with wonderful stories about them, and immediately the breed became popular. Police dogs were imported by the hundred from Europe, and hundreds of breeding kennels were started here in America. Presently there were fully a dozen times as many police dogs in the United States as there had been a year or two earlier. For a time they were the most popular of all breeds. Fanciers went daft about them. Soon the market was glutted with police dogs. That kind of sudden popularity is not good for dog or for human. It is like the fad for a certain kind of dress or suit or hat that everybody begins wearing. It dies out nearly as suddenly as it begins.

But the people who really understood these dogs were not affected by changing canine fashions. They kept on with their dogs, paying no attention to the passing of the craze. So the police dog remained in great numbers. Every year more and more people appreciated him for what he was. He was too great a dog for fashion to spoil.

# The Setter

ONCE when I was a boy, my father and I were quail-shooting with my father's staunch old setter, Frank. The dog was in front of us quartering the stubble field, working away happily, with tail waving and nose almost touching the ground. A railroad track ran through the middle of the big field. As Frank came within about eighteen inches of the track, he halted and stiffened all over, with one forefoot tucked up. He was pointing something in a thick tuft of grass just across the track, and waiting for my father to order him to go ahead. A good hunting dog does not break his point without orders, and old Frank was the very best.

Just then an express train came tearing around a curve at top speed. For a moment I thought the locomotive was going to hit the dog. He must have thought so, too, for it came thundering toward him in a swirl of dust and smoke and cinders. Such a sight was enough to scare any dog and make him back away. But Frank did not stir. When the train had whizzed past, missing his rigid nose by perhaps only an inch or so, filling his sensitive nostrils with cinders and black smoke, and stinging him with its shower of hot sparks, there he stood motionless as a statue, his forefoot tucked up, body and tail rigid. He still pointed the unseen occupant of the patch of grass, and waited for my father's command.

At a word from my father, Frank took a hesitant step forward, then another. Out from the tuft of long grass a quail flew up. I don't know why it had not flown up when the train passed; perhaps it was too badly frightened. As it rocketed from the hiding place whence his slow advance had driven it, Frank dropped to the ground, as he had been taught to do when a bird broke cover.

My father shot the fast-flying quail; then gave Frank another order. Instantly the setter dashed forward and retrieved the fallen bird, carrying it as tenderly in his mouth as if it had been a cushion full of pins. He trotted back to my father and dropped the quail gently into the wide-open pocket of his hunting coat. He seemed to take great pride in this simple trick.

That is all there is to the story. I have told it because it shows several things. First, the perfect training and the courage that kept Frank at his post of duty even when death seemed to be thundering down upon him; then the skill at retrieving; and, most of all, the unerring power of scent which made him locate the unseen bird and which enabled him to keep at the point when cinders and smoke were almost suffocating him.

There is nothing remarkable, perhaps, in Frank holding his point under such dismaying conditions. Many another perfectly trained setter might have done the same. But here is the remarkable thing about him, and about any setter, or indeed about any good hunting dog, a thing which you may not have stopped to consider. In the field, every setter is acting upon instincts that go back for countless centuries—instincts for using his marvelous powers of scent to find and to track down game. Originally, the hunting dog did not do this for his master. He had no master. He did it for himself, in order to get food to keep him alive. That was his one way of getting a living. It was his strongest instinct and desire.

Yet the hunting dog has subdued that instinct and desire, and has made it serve mankind and not himself. All day and every day, he will hunt game, of which he knows he is not going to receive a single mouthful. His work and skill are all for the benefit of his master. Often the dog enjoys the hunt, but there is no profit in it for him. Its proceeds all go to man. He works without pay or hope of pay. Is there any finer proof of unselfish devotion that he could give?

There are several kinds of setters: the beautifully shining auburn dog, known as the Irish or red setter; the black-and-tan Gordon setter; the ticked or gold-and-white English setter; the Llewellyn, and others. But they all come from the same original stock; and all are true setters. Sometimes they are crossed with pointers, and then we find the grand hunting dog known in the South as the dropper.

In olden times, so it is said, setters were used by huntsmen who netted game instead of shooting it. The dog would go ahead until he scented a hidden bird, and then would point it. The men behind him used to creep up silently, carrying a large net. At a signal, the dog would crouch low, or "set," in order that they might throw the net beyond him and over the bird. Hence the name setter. I do not vouch for the truth of this theory of the origin of the setter's name; but it is the most generally accepted.

It is believed, too, that the setter is what is known as a "made" breed. That is, in the beginning it was not a regular breed, but was evolved from some other. The setter is believed to have been evolved, many centuries ago, from the hunting spaniel. Sportsmen decided that the spaniel would be improved as a hunter if certain changes could be made in his size and build. So, with this aim, they bred certain types of spaniels generation after generation, until at last they evolved an animal closely resembling the present-day setter.

It seems quite likely that this is true, for no other breed of dog lends itself so readily to changes of all kinds as does the spaniel. He has more variations of size and shape than any other established breed. For instance, in the time of Henry III of France, spaniels were bred so small that one of them could be carried in a muff; and, at the same time, among hunting spaniels, they were bred to weigh more than forty pounds. That seems impossible. Yet, even today, consider the vast difference in size and looks between the tiny King Charles spaniel and the rough-and-ready Clumber spaniel. Both sprang from the same ancestry. If you will note the long, silky ears of the setter, his texture of coat, and his general expression, you will see a strong resemblance to a spaniel, in spite of the difference in his general build. I wonder nobody seems to have thought of evolving a miniature setter to be carried to the hunting field in a pocket or a satchel!

Apart from his wondrous value as a hunter, the setter has another use which is almost as important. He is a grand house dog and pal and guard. He adapts himself to the ways of the home and of the household as do few other dogs. Some of the finest pals I have had were setters. My favorite of all these was a gold-and-red setter — cross between English and Irish — named Duke. He was my chum for six years and I was the only member of the family he made any pretense of obeying.

Once a pet white rabbit belonging to the children got out of its hutch and ran off to the woods. The bunny's little owners were inconsolable. Duke hated to see the children unhappy. He seemed to understand perfectly what was the matter with the youngsters. And it made him break one of the most rigorous laws laid down for every setter or pointer — the law that they must scent only birds, and must pay no attention at all to rabbits. Sportsmen often cure young setters of pointing rabbits or chasing them, by beating them unmercifully with the body of the dead rabbit.

Duke vanished while the children's lamentations were at their height. Nobody saw him go. But everyone saw him come back. In less than ten minutes he came loping toward the house from the nearby woods. In his mouth he was carrying something white that kicked and squirmed vainly to escape. Carefully he laid the rabbit down on the veranda floor among the delighted children; then turned shamefacedly away, as if knowing he had transgressed the most sacred law of the field dog.

The rabbit was not hurt or even pinched or bruised by that chase and capture and quarter-mile journey in Duke's mouth. A non-hunting dog would have mauled it terribly. Never before or afterward did Duke pay the slightest attention to the bunny. And never afterward did the rabbit get loose or run away from its young owners. If it had, I am certain Duke would have brought it back again.

Not all sporting dogs or working dogs make ideal house dogs. But nineteen setters out of twenty are as perfect in the home as they are in the field, if sense and care have been used in their training. Dogs are like children, in one way: the more sensitive and highly developed and clever they are, the easier it is to spoil them by injudicious early training. On the other hand, the easier it is to develop their characters and brains by the right sort of upbringing. A thousand-dollar watch is easier to wreck than a two-dollar clock. But, rightly treated, it gives you a thousand times as much satisfaction. Do you get the idea? It will be well for you to remember it, in case you have the job of training a high-bred dog.

Here is another instance of a setter's brilliant mind and independent character.

Koos was an elderly Gordon setter belonging to a neighbor of mine who lived near Sunnybank. Koos, by the way, is the Kaffir or Zulu word for chief. He was a mighty hunter; and he took keen pride in his field prowess. Sometimes his owner would lend him to a friend who wanted to go out for a day's shooting.

Koos was glad enough to accompany an outsider. He was so fond of hunting that it made little difference to him who the gunner might be, as long as he was a good sportsman and a crack shot. What Koos wanted was to find and point and put up a goodly quantity of birds; and then to have those birds shot and to retrieve them. Nothing else mattered much to him.

One day he was lent to a young man who was a more zealous than accurate shot. Half an hour after they started on their hunt, a covey of quail was flushed. The gunner blazed away with both barrels. The scattered covey of birds sailed off unhurt. Koos looked at the man reproachfully, but continued to quarter the field in quest of the various members of the covey.

Soon he pointed one of them. The gunner drew near, and fired as the bird whizzed upward. Again he scored a clean miss. Again Koos glowered reproachfully at him, hesitated for a moment, and went on with his hunting. A third quail from the scattered covey soared upward within easy range. The hunter shot both barrels at it. For the third time in succession he missed.

That was enough for Koos. The old dog turned around and trotted toward home. Nor could the hunter's shouts and coaxings lure him to so much as look back at him. The dog did not stop until he reached his master's yard. There he went into his kennel and lay down. He did not care to associate longer with such a bungler. Do you blame him?

# The Scottish Terrier

HE HAD a hard time getting a name, did the Scottish terrier; and the dispute split the dog fanciers of Scotland and northern England into two fiercely arguing factions. This was more than half a century ago.

In those days, half a dozen different kinds of terriers bred in Scotland went under the general name of Scotch (not Scottish) terriers. I have seen pictures of these old-time Scotch terriers. Only a few of them looked alike; and none of them looked like the good little dog known to us as the Scottish terrier. Then from the far northwest corner of Scotland, several sturdy terriers of an unfamiliar kind were brought down to the British dog shows. Some experts said the newcomers were just a variety of Highland terriers. Others said the new dogs were Aberdeen terriers. That was how the squabble began.

But, by whatever name he was called, the little dog from northwest Scotland began to attract attention. He was growing popular everywhere. And still the dispute raged as to whether to call him a Highland or an Aberdeen. He was loved, but he was nameless. Or, rather, he had two names; and neither of them fitted him.

Then some peacemaker, whose name I don't remember, hit on a clever plan for reconciling the two warring factions and for giving the new dog a name. At a kennel club meeting where the quarrel once more waxed hot, he said: "It's all Scotland, anyway, whether you use the name Highland or Aberdeen. Why not compromise and call him the Scottish terrier? You can't possibly go wrong on that."

Both sides agreed. They and all the public were glad to get an authoritative name for the popular late arrival in dogdom. So he became formally the Scottish terrier. The Scotch are great folk for nicknames and for shortening a name that is long. That is why they call the Shetland pony a Sheltie. That is why they shortened the words, Scottish terrier, into Scottie.

But there was more to be done before the fanciers were satisfied. There were many arguments as to what the true type of Scottie should look like. For instance, whether he should have long, folding ears like a Sealyham terrier, or tulip-tipped ears like a collie; or whether his ears should prick straight upward. At last the victory was won by the fanciers who thought he ought to have prick-ears. And after that all Scotties were bred with the aim of making them have ears that stood straight up. That is why your Scottie, today, has stand-up, pointed ears.

When all these points were settled, the Scottie began his present-day career. He had a name; and he had the several arbitrary points which the show-ring demands. He was complete. Moreover, he had come to stay. Every year he was more and more popular. One reason for his hold on the public heart was that he was not only a man's dog, but a woman's dog and a child's dog, as well. There was nothing of the shivering lap dog about him. He was not a toy. Yet along with his splendid courage and hardihood, he had an underlying sweetness and gentleness and a whimsical sense of fun that made him an ideal house pet. In short, he was one of those rare finds, "a dog for the whole family."

I have a warm place in my heart for Scotties because of a few of my own experiences with them. I want to tell you about my friendship for one Scottie, in particular. Perhaps you will see why I grew so fond of him.

His name was Roddy. He belonged to my sister, whose summer home adjoins Sunnybank. Roddy was small, even for a Scottie. But his tiny body held the heart of a lion. Sometimes one or two of my biggest collies would follow me when I went to see my sister. As soon as these giants crossed the boundary line into her land, Roddy was on the job as an official watchdog.

He would rush out of the house, bristling with rage, and dash at the collies. Straight up to them he would charge, intent on driving them from the premises. He was a wise little dog, and he must have realized that any one of the collies could have killed him with a single bite, but it was his duty to defend his owner's land from marauders.

If the Sunnybank collies had been dogs of certain other breeds, it might have gone hard with valiant Roddy. But my experience has been that not one collie in five hundred will harm a dog so much smaller than himself.

As Roddy leaped up in a vain effort to reach their throats, the Sunnybank dogs would stand looking amusedly down at him, waving their plumed tails. From one to another of them he would rush, challenging each to mortal combat. Then, when they would neither fight nor run away, he would glare at them in utter disgust, and trot angrily back to the house. It was no use. He had done his best to be a murderously dangerous watchdog.

He felt as you or I might have, if we had dared the Statue of Liberty to step down from her pedestal and fight us, and she had merely smiled and stayed where she was.

But Roddy's courage was as flamingly bright as if they actually had attacked him. He was on guard of his owner's home. And he was as fearless a guard as if he had fifty times his size and strength. That was proven one hot summer day.

An enormous ice-wagon lumbered clumsily into the driveway leading to my sister's house. Roddy seemed to regard this vast intruder as a worse enemy to his owner's premises than were my collies. So, head down, he charged it. Luckily the unwieldy vehicle had nearly passed before he could get to it. However, he could still bite the nearer rear wheel. Instead, it passed over Roddy's nose, just in front of the eyes, crushing the bones and flesh to a hideous pulp.

The little dog was picked up and carried tenderly indoors. He was suffering tortures. But he did not cry out or even whimper. The pain could not wring a sound from this hero. He was put into a soft basket. Gravely he stood there, statuelike, giving no faintest sign of his agony.

My sister telephoned to me. I gave her the addresses of the two best veterinaries in the region. While she was telephoning for them to come at once, I arrived at her house. My good friend Roddy looked up as I came in and made a feeble effort to wag his tail. As his dark eyes met mine, I had a strange sensation that he was appealing to me for help; and that he was too proud to give voice to such appeal. Perhaps that was an idiotic fancy, but it gave me a resolve that Roddy was going to be cured if the finest skill and nursing could do it. I examined the smashed nose; and I had little hope. But he and I were going to see it through, together.

Then both vets arrived, almost at the same minute. They made a careful examination. After the inspection was over, they conferred together in whispers. One of them turned to me and said:

"There is no possible chance for him. Both jaws are broken to pieces. He can't live for a single day. Tetanus or other blood-poison would set in. The only thing to do is to put him to death now, as mercifully as we can."

"If my jaws were smashed to atoms," I demanded, "would a doctor decide I must be put to death at once? I think not. He'd find himself in court for murder, if he did. And you men aren't going to kill Roddy. You're going to set that jaw. Both jaws. I don't know how you're going to do it, but you are. And I'm going to stand here with my back to the door, to see you don't go till you've finished the job."

Yes, it was a bullying way to talk. I realize that now. But I couldn't forget the queer look of trust and appeal in Roddy's eyes when he saw me. At last, after many refusals, both vets set grumblingly to work. I stood for three hours, watching them, and now and then talking to Roddy. It was not a happy three hours. But if Roddy could stand it, I could.

At last the work was finished. Mighty clever work it was, too. In spite of my ignorance of surgery, I could see that. Roddy's nose, from the eyes forward, was bound fast in splints and plaster, with small openings at the nostrils for him to breathe through.

"Well, we've done it," announced one of the vets. "But it won't do any good. He will die of blood-poisoning inside of three days. Besides, how is he going to take any nourishment?"

"Bore a small hole in the front of that plaster mask, just between the lips," I told him. "A hole big enough for a straw or a glass tube to go through. We can feed him milk and beaten eggs, a spoonful at a time, through the tube."

Two months later, Roddy was as well as ever he had been, except for the loss of three teeth that had been crushed under the wheel. His jaws were perfectly sound. He could chew as well as ever. Not only that, but he could bite just as ferociously and just as uselessly at the furry throats of my collies when they strayed into his owner's grounds. He lived six years after that. I have always believed his gallant heart and his grit kept him alive when both doctors said he must die.

I think the story which I have just told gives as full an insight into Scottie character as anything that has been written of the great little breed. I have had other Scottie friends, too, besides Roddy; and the better I know such dogs, the more I admire and like them.

In olden days the Scottie's ancestors were not kept merely as pets. In the Highlands they had work to do — hard and dangerous work at that, work from which a lesser dog would have flinched. Perhaps that is how the Scottie got his fearless heart.

When a pack of fox-hounds or otter-hounds were taken out for a day's hunting, along went a Scottie. Sometimes he jogged among the pack on his own short legs, but when there was fast going, he was carried across the saddle-bow of one of the huntsmen. When the fox or otter was run to earth — that is, when he had found some hole in the ground where he could take refuge — the big hounds danced helplessly around the hole, barking at the top of their lungs, unable to get at their prey. As far as they were concerned, the hunt was over.

This was the time when the little terrier did his work. The name terrier comes from the French word *terre*, meaning earth; a terrier was a dog that worked underground. The Scottie would dig at the hole until it was large enough for him to wriggle in. Down he would go, until he came to the fox or otter hiding there. A sound of snarling would be heard by those above, witnessing the fact that a life-and-death battle was going on underground. At last, the sound of subterranean warfare would die down. Then either the terrier did not come out at all, or he would come backing slowly out of the hole, dragging his dead enemy out into the open to lay it at the feet of his master. True, the Scottie might be scarred and bleeding, but his tough little body soon healed, and he was good for another battle.

But if the Scottie did not come out, and spades were brought, and men dug down to see what had happened, almost always they found the otter or the fox also lying dead alongside the terrier it had killed. The Scottie had won his fight, even if death went along with victory.

I wonder that such a life did not make him and his descendants savage and mean. But savagery and meanness have no place at all in the make-up of the true Scottie.

There is another thing about the Scottie that makes him valuable nowadays. If he is given a good amount of exercise, he seems to thrive nearly as well in a city apartment as out in the country. He is one of the few *real* dogs that seems at home in the city. I have known of Scotties that lived for many years in apartments, with no exercise outdoors except a daily walk of a mile or so on a leash, along hard pavements. Perhaps it is their native hardiness that enables them to do this. Personally I won't keep a dog in a large city. But there are many thousands of people who have to live in cities for all or almost all of the year, and yet who crave a dog as comrade and guard and pet. For these people the Scottie solves the problem of dog-owning and city-dwelling.

I like the Scottie. I have just been trying to make you like him, too. Have I succeeded? I like to think so. For he is well worth your friendship.

# The Cocker Spaniel

THERE are more kinds of spaniels than I have the time or the memory to tell you: the cocker, the Norfolk, the King Charles, the Springer, the Clumber, the Sussex, the Irish water spaniel, the field spaniel, and so through the endless list. Each of them is good — very good indeed, and each is popular with certain groups of fanciers.

But of all the many spaniels, the cocker is more popular than any of the rest, perhaps more popular than all the rest put together. And he has stayed in favor for a longer time without showing any sign of losing his tremendous popularity. Queerly enough, the more famous he has become, the smaller he has become in size.

Something about the spaniel's make-up seems to make it easier to change him in stature and shape and character than other dogs. Sportsmen and fanciers have taken advantage of this, to do all kinds of things to him. Some spaniels, for instance, are eight times as large and heavy as others. Some are frankly "toys" and are guarded from cold and fatigue as carefully as if they were newborn human babies. Others are huge and powerful, and are used to all weathers and all labors. Some are tiny lap dogs. Some are mighty hunters. But, as I said, the cocker is the best-loved and most famous of them all.

Besides, the cocker is one of the very few spaniels which is not only an ideal house pet but also a first-rate gun dog. He seems to combine all the best traits of the whole spaniel tribe, from largest to smallest. Few people nowadays use him for hunting, but he can be trained to do it with ease. In fact, that is said to be how he got his name. He was used in Great Britain to hunt moor cocks, woodcocks, and the like; and for this reason was at first called a cocking spaniel. The name was shortened to cocker; and as time went on, he was kept more and more as a pet, and less and less as a field dog.

When this change came, breeders began to evolve smaller cocker spaniels. The earlier dogs of the variety were more than a third larger than those you see today. The cocker became so well liked in England that a few American fanciers imported him. Then the contest began between the British- and Yankee-bred cockers in the bench shows on both sides of the ocean. The top weight-limit was set at twenty-eight pounds.

Steadily the American cockers drew ahead of their British brethren in popularity. Ever since 1881 their vogue has been growing. Up to that year, there were few of them in this country; and most of those would be hard to recognize as members of the variety we see today.

Then came the effort to breed the cocker in as many colors as possible. Presently they were seen at the shows with jet-black coats, with "ticked" coats, with brown or red coats, with cream-colored coats. Cockers of one coloring are probably just as good as are cockers of any other. It is all a matter of the buyer's taste.

I have been telling you about the appearance of the cocker, and something of his history. But the best part of any dog is the part he thinks with, the brain or heart that makes him wise or lovable.

Queen was a cocker spaniel of no specially lofty pedigree. She belonged to a grocer in Columbus, Ohio. Her owner had the sense to understand her nature and to give her an education, which she was quick to absorb. One of the first things he taught her was to mail his letters. Many dogs will retrieve their owners' mail from the carriers, but few have been taught to post it. Queen did it perfectly, and always took the keenest pride and delight in acting as mail-carrier.

The mail was put in a basket. Around Queen's neck was slung a little placard which read: *Please mail my letters in this box.* The cocker took the basket in her mouth and trotted away to the nearest mail-box. This was on a lamp-post at a busy corner, nearly a half-mile from the grocery. There, with the basket still in her mouth and the placard around her neck, she would sit up on her hind legs, and wag her tail invitingly at every passer-by who stopped to look at her.

Within a minute or two, someone would be certain to take the bundle of mail out of the basket and drop it into the box. Then, wagging her tail still more briskly in gratitude, Queen would run home with the empty basket. In all the hundreds of times she performed this trick, never was a letter lost.

Her master also taught her to know the names and the addresses of the few tradesmen with whom he dealt. He would give her a basket and tell her to go to his butcher. Not once would she go to the wrong place. Into the butcher shop she would run with her basket. The butcher would read the order list lying in the bottom of it, and place the provisions in the basket for Queen to carry safely back home. The same thing happened at other stores.

No matter how hungry she might be, and no matter how tempting the food in the basket might smell, she would not touch a mouthful of it; and she fought fiercely to defend it whenever some other dog or some man tried to rob her of it.

Queen had one enemy. This was Rosie, the grocer's cat. She and Rosie hated each other; but the cocker obeyed her master's orders and let Rosie alone. One day while she was nosing around the grocery cellar, Queen came upon a box with four kittens in it. The kittens belonged to Rosie and they were about ten days old.

For once, Queen forgot her master's command. Ferociously she drove the mother cat away from the babies. Then she herself settled down among the kittens, cuddling them close to her, licking them all over and crooning to them. From that time on, they belonged to her.

Except when the grocer shut her up in another room, several times a day, so that Rosie might come back to her babies, Queen spent all her time crouching there among them, driving the mother away. She had adopted that litter; and she continued to brood over them and to wash them and to guard them until they were able to look after themselves. She never had any children of her own, and all her baffled mother-love went forth to Rosie's kittens.

If you have a cocker spaniel, you have a grand little dog. Make the most of him; and treat him well. He deserves it.

I am going to tell you another story to show you another side of the cocker's nature. Jim was a cocker that belonged to a rancher's son, near Loveland, Colorado. The boy taught Jim to hunt rabbits. Jim took to the job as delightedly as Queen had taken to Rosie's kittens. He loved to hunt. The only trouble was that he would go coursing rabbits on his own account, when his young owner wasn't along. And that got him into a lot of trouble, as you shall see.

He went out into the woods one day alone, on such a hunt. He "put up" a rabbit in a clump of bushes, and gave chase. The rabbit dived into a small irrigation pipe that led from a ditch and a far-off reservoir. Jim dived into the pipe in hot pursuit. At a sharp angle in the pipe, he caught and killed his prey.

Then the unfortunate dog found he was jammed into such a narrow corner of the piping that he could not go forward or back. He was stuck underground, and in danger of staying there until he should starve to death. Nobody knew where to look for him.

Many a dog would have writhed and howled till he wore himself out. Jim was not an ordinary dog. With all his might he thrust upward, using the top of his head as a battering ram. He continued to hammer as fast and as hard as he could, until his head was dizzy and bleeding. By rare good luck, a man was on his way home through the woods just then, and halted at the sound of the thumping beneath his feet. He dug down and uncoupled a length of piping at the corner where Jim was stuck. He lifted out the gallant spaniel whose life had been saved by his refusal to give up the fight.

Do you wonder I said that a cocker is a grand little dog?

# The Husky

HE HAS several names: husky, Malemute, Alaskan dog, Eskimo dog, and one or two more. But husky and Malemute are the names most used. Originally, the Malemute was supposed to have more wolf blood in him than the husky. All of them are descended from a cross between the fierce and powerful wolf of the northland and the domestic dog.

In the far north a team of horses would be of no use at all in carrying loads through mountain-deep snow or over thin crust or across miles of glaring, uncertain ice. The average dog was not fitted for such work. True, dogs have been drawing carts in Europe for centuries. But there is a world of difference between pulling a small, wheeled vehicle over fairly smooth ground and forcing a way through gigantic drifts in a blizzard when the thermometer is perhaps fifty degrees below zero. No, the ordinary dog was not geared for such work. The wolf was splendidly suited for it. But the wolf would not work. He could not be tamed and he could not be trained. The dog was willing to do it, but couldn't. The wolf was able to do it, but wouldn't.

The hard problem solved itself. Female work dogs in the Eskimo villages had litters of pups whose sires were wild wolves. Some of these crossbreed pups were allowed to live and to grow up. It was found that the best of them had all the tireless endurance of the wolf, along with the teachableness of the dog. That was the beginning. A new breed was evolved, or, rather, not a true breed, but a hybrid later to be known as the husky.

By the time the first white men came to the north, they found the breed long established—wolf-dogs that could draw sledges for hundreds of miles, over ice and snow, steep mountain and deep gorge.

The animals could go long without food or drink—as wolves needs must do in their wild state — and throve on a cheap diet of dried fish and such food scraps as they could get. They were the only draught-beasts that were not to be baffled by the wildest weather and the toughest going.

This breed was not evolved and developed, as are most others, for chumship or for sport. It was born of stark necessity for means of travel. Its sole mission in life was to work until it should die or be killed. Grim brutes, bred for grim needs. These dogs opened up the northland to trade and to civilization.

Most assuredly, the husky was not a pet. Fear or a sense of duty, or both, held him to his work. But he had too much of his wolf ancestor's cruel savageness to be a true pal to mankind. One or two old-time Alaskan mushers have told me that once in a thousand times they have found a husky that was a good chum and seemed to feel a genuine devotion toward his master. But these cases, they said, were so rare as to prove nothing. He did his work pluckily and well for the most part. But he would not accept man as a comrade or as something to worship. He remained fierce. Usually he remained treacherous, too. Even when he was brought up from puppyhood far from his native north and among gentle surroundings, it was seldom he forgot his mixture of wolf blood.

I used the word musher just now. It was the name given long ago to drivers of dog-sleds. When the French and the French-Canadian hunters and trappers and traders first came to the north, they spoke to their huskies in French. When they wanted a dog-team to start going, they shouted: "*Marchez!*" or "*Marche!*" Englishmen and Americans who came later to Alaska or to other parts of the north, corrupted this to "Mush!" That is the word still in use in driving dog-teams up there. And the dog men are still known as mushers.

The best-known husky in all the dramatic history of the north was the great Balto. A statue to his memory has been erected in Central Park, New York. It is perhaps the only statue on any public property in the United States dedicated solely to a dog. If there is such another, I never have heard of it. But Balto was well worth that high honor. Some of you may have read the story of his magnificent heroism and of his mighty service to suffering mankind, but the tale is well worth retelling; even as is the narrative of any heroic deed.

Nome, Alaska, was cut off from the rest of the world by a terrific winter storm which raged unceasingly for many days. Except for telephone and telegraph, the city was as isolated as any desert island. And it was far more impossible to get to or to get away from than a desert island.

It was at this time that a diphtheria epidemic broke out virulently in Nome. No anti-diphtheria serum was on hand. Without the use of this preventive and cure, the epidemic seemed likely to spread, and thousands of lives might be lost. Seldom in modern times has a civilized community been in such dire peril.

Serum had to be brought at once. But how was it to be done? The storm was so devastating and the going so hard that there seemed no chance for even the strongest dog-team to get through to the stricken city. Yet a musher offered to make the effort.

With a goodly supply of serum, he set out for Nome. The whizzing snow made everything invisible, except for a yard or so ahead of the dog-team. The brave huskies battled through it as best they could. But they made scant progress. Even the veteran lead-dog — the dog that is foremost in the line when the team is hitched — faltered and seemed uncertain of his way.

The driver could not be wholly certain his team had not strayed from the unseen track. On every side were ice-fissures or gorges which might engulf him and his dogs and sled, along with the precious serum. Blindly he staggered on, often halting the team and going ahead to stamp down snowdrifts so the dogs could draw the sled over them. But the lead-dog grew more and more uncertain. Others in the team were scared or exhausted. The chance of reaching Nome was all but nothing.

One member of the team was a liver-colored husky named Balto. He had a chest as wide as that of two average strong dogs, and legs as thick as those of a yearling calf. He had unbelievable strength and willingness; and he was one of the few sled dogs in the north that loved his master and had no trace of wolf savagery in his great heart. And now it was to Balto that the driver turned as a last hope.

The man unhitched the team as fast as his numb fingers could achieve the task. Then he hitched it again with Balto in the lead. The sled dogs had dropped to the ground in utter fatigue the moment they were halted. Some of them lay panting and helpless. A few began to wriggle down deep into the snow for warmth; this being the habit of sled dogs when night camp is made.

But Balto would have no shirking. Nobody could doubt that he knew what responsibility was now his. There was to be no more lying down in harness, no slacking or hanging back. Balto was on the job; and Balto was going to keep the others on it, too. He had been a submissive follower. Now he showed himself an inspired leader. Moreover, he was able to put some of his enthusiasm into the rest of the team. Urged by some mystic sense of direction, Balto kept to the track. By an instinct which could not have been mere luck, he steered his mates away from the brink of chasms and skirted the treacherous air-holes in the ice. Firmly he checked any tendency of the other dogs to slow down when exhaustion crushed them.

At last, half-dead, the dog-team and its driver staggered into Nome with the anti-diphtheria serum (or antitoxin, if you like that term better) which was to save so many human lives. The whole world rang with the grand lead-dog's praises. Balto became a national hero. Nothing was too good for him. He had rescued Nome.

It was one of the wildest and most perilous journeys in history. That it did not end in disaster instead of glorious triumph was due to Balto. He guided where none but he could guide. By his own fiery strength and courage and example, he made his team-mates achieve the impossible. Don't you think he deserved a statue?

It would be pleasant to say that the great dog was allowed to live out his days in peace and comfort, as the loved comrade of his master and in the region where he had been born and which was his home. But the story has no such happy ending as that. I wish it had. Let me tell you the rest of it.

It was decided to make money out of Balto's world-wide fame. So he was shipped to New York to be carried around from one vaudeville show to another. Every day, for a while, he was taken out to Englewood, New Jersey, where he had to pose for the statue which was to stand in Central Park — the statue to be erected in honor of a hero.

It was at Englewood that I met him and made friends with him. I used to go to the sculptor's studio while the modeling of the statue was in progress. Once or twice, during the rests, I took Balto for a walk. This was more of an athletic ordeal than a pleasure for me. For the dog would throw his huge shoulders against the breast-strap of his harness, clamp his furry tail up over his back, and set off at a pace I could hardly follow. All my strength was needed on the leash to keep him from carrying me off my feet. Yet, in spite of this, I got a certain enjoyment out of the walks; and Balto got much more. He would go back to the studio with me, refreshed and ready for another long pose on the sculptor's "throne." These long poses must have tired and annoyed the lively dog. But he bore them with grave gentleness, never fidgeting.

With the same quickness of perception and with the same steady sense of duty as had made him fit to lead the dash to Nome, the splendid husky learned what was expected of him at the studio. He took the position desired, and stood motionless while the sculptor worked. Never have I seen better posing by a model.

The studio and the Englewood walks were pleasant enough, I suppose. But the vaudeville work could have been nothing but torment for the dog whose life had been spent in the cool silences and illimitable open country of the north. Day after day, in the hot city, he had to appear on a dusty stage before crowds of spectators; his ears assailed by blaring jazz music and chatter; his eyes itching from the dazzle of artificial lights; his sensitive nostrils teased by alien scents.

Day after day and night after night, for week after week, it went on; the husky's only walks in the stuffy alleyways behind the theaters; his sleeping quarters all but airless; his life turned into a never-ending exhibition. Think what it must have meant to Balto after the northland he loved!

He bore it all calmly and gallantly, as he had borne the posing at the studio. But I wish I could have bought him, just then, and turned him loose here at Sunnybank or could have sent him back to the north. Seldom have I been so sorry for any one or for any thing. Yet I was helpless to free him. He was making too much money for his handlers to be allowed to quit the show. That was his reward for risking his life for mankind on the blind trail to Nome. I don't know what became of him, after his show days were over; or if he survived the ordeal and was at last taken home again.

The huskies have been put to another use in the Arctic regions. Packs of them have been trained for hunting. They have been taught to trail and run down the musk ox; to hamstring him and to hold him helpless until their masters can catch up with them. They have learned to follow the polar bear, as well; and to surround him and hold him cleverly at bay while the hunters are getting within gunshot distance of the snowy prey. They are fearless and crafty, these bear and ox trailers. The wolf instinct teaches them more wise tricks in hunting such big game as the north affords than any mere dog could possibly know. For their ancestral wolves had to make their living that way for hundreds of years; and the teachings of those wolves have been carried on through their descendants.

That same wolf instinct makes the husky a clever fisherman, too. The average dog can't catch a lively fish; though many dogs of mine have spent hours in the shallower parts of our lake here, in an effort to seize minnows. But the husky knows how to do it. He catches fish by the dozen — good big fish, at that — in the pools and shallows of Alaskan rivers.

People are prone to think of Alaska and other parts of the far north as bitterly cold for twelve months of the year. This is not so. The summer is short, but it is hot. Sometimes the heat is as intense as in our own part of the world. And during that hot season, the mosquitoes and black flies make life a burden to everyone.

The huskies not only can stand the cold at sixty degrees below zero, but they can thrive in the choking heat of summer with its million insect stings. They can do even more. They can pick up their own living at such times; and "live off the country" as easily as does a wolf. This they have to do, many of them. For there is no sled work in summer; so usually there is no job for the sled dogs. Many of their owners turn them adrift to forage for themselves in the wilderness, until the early return of winter makes them useful again for carrying mail and provisions from one scattered settlement to another. They act as village scavengers in the warm off-season and range the forests for game.

The husky is the best and most necessary work dog the world has known. But for him, the lines of communication in the north would be clogged for weeks at a time. True, as a rule he is not exactly lovable or companionable. But then, neither is a freight locomotive. And the one is as highly useful as is the other.

At some of the big dog shows, nowadays, a team or two of huskies is exhibited. Next time there is such a team shown in or near your home town, go to see it. But take my tip, and keep out of biting distance of the surly huskies. Admire them. But don't pat them.

# The Boston Terrier

THE label, *Made in America*, belongs to the Boston terrier more than to any other breed of dog. For the Boston is one hundred per cent American. He was evolved here; and here his build and disposition were perfected. He is much liked in Europe, too, of late years. But he is all Yankee.

For that matter, isn't he the only dog with an all-American name? If Uncle Sam should ever decide on another emblem in addition to the American eagle, he could do worse than pick out the patriotic little Boston terrier. For the Boston belongs here every bit as much as does the bald eagle. Yes, his ancestors originally came from Europe. So did yours and mine. And these ancestors of his were not Boston terriers. At the time they crossed the Atlantic, there was no such dog as the Boston. But there were bulldogs, and there were terriers of various kinds. Those were his forebears.

In earlier days, the cruel sport of dog-fighting was much more popular than it is now. The bulldog was a lion-hearted warrior. But he lacked speed. There were several kinds of terriers which also had the hearts of lions and were as quick as cats and were born fighters. But they lacked the bulldog's heavy strength. So a blend was made, by sporting men, of bulldog and terrier; in order to get the best battling qualities of each.

Two or three distinct types of dogs came from this crossbreeding, the most favored fighter of them all being what is called the pit terrier. He was a murderous fighting machine, and could lick his weight in tigers. He loved nothing as much as to fight. It was his nature.

Back in the 1880's, one of these pit terriers, Judge by name, was bought by R. C. Hooper of Boston from William O'Brien, another

Bostonian. Judge was dark brindle in color, with a white blaze on his face, and he weighed a little more than thirty pounds. He was a famous fighter. His pedigree is unknown. He was mated to another Massachusetts pit terrier, named Gyp — whose pedigree is likewise unknown, but who also had been bred for fighting. From this mating, the entire race of Boston terriers is descended. Judge and Gyp were the Adam and Eve of the Bostons.

Their pups were of a distinct type, different from the average pit bull-terriers, and so handsome that the strain and general appearance were purposely continued and intensified in later breedings. Presently there was a distinct type of pit dog along these lines, but it was known as a strain and not as a breed.

Owners of these dogs formed a club. They called it the American Bull-terrier Club, and its object was to perfect and popularize the new strain they had developed. But the American Kennel Club would not consent to their using this name, on the plea that the newly developed dog was no longer a bull-terrier. James Watson, then an official of the American Kennel Club, wrote later of this application:

"We suggested that as their dog was not a bull-terrier at all and was bred only in Boston, that it would be better for them to take the name of Boston Terrier Club. This was in 1891. The American Kennel Club, however, did not admit the Boston applicants until 1893."

So you see there was no such name as Boston terrier until 1891; and it does not seem to have been formally recognized as a breed by itself until 1893. By the latter year, the new dogs had made tremendous strides toward popularity. They were bred smaller and of more graceful shape than at first; and no longer were they used as professional fighters. They had become almost exactly like today's Bostons.

It seems odd that the present-day Boston terrier is so gentle and friendly, when you remember that up to a very few decades ago his ancestors were bred only for fighting. The gallant spirit of those old-timers remains in their Boston descendants, making them brave and independent. But no trace is left of savagery or of the craving to murder every other dog they can get at.

There is an old saying that there are finer ways of showing courage than by fighting. Here is a true story of a Boston terrier. I will make it short; even as its hero's life was short.

Micky lived in southern Florida some years ago. His blood may not have been as blue as that of bench-show Bostons. But his pluck was as great as that of the pit terriers who were his ancestors. And his heart and brain were those of a true pal and playfellow.

Micky was the loved friend of the town's school children. He romped with them at recess, shared their lunches and went swimming with them after school. He was the favorite dog in town. His gentleness and courage and gay sense of fun were enough to make him popular with youngsters and grown-ups alike.

He took upon himself a duty which nobody had taught him. Every day he would escort a group of the tiniest children in the neighborhood to and from school, waiting on the steps for them to come out, and stalking belligerently between them and anything or anybody he thought might menace them. It was a pretty sight, this twice-a-day parade of which he was grand marshal. Parents felt safe about their children when Micky was along.

One day a six-year-old boy was very late in starting for the primary school. Micky was trotting back from his regular escort duty when he caught sight of the hurrying child. Instantly he joined the tardy little fellow and accompanied him on his way. To make up for lost time, the boy ran across a sandy vacant lot, as a short cut to the schoolhouse.

Suddenly Micky darted forward. He had seen what the youngster had not — a big rattlesnake lying directly in the path. At the sound of their approach, the snake lifted its ugly three-cornered head and drew back to strike, its rattles sounding a harsh warning that the child did not notice. Instinct told Micky the hideous fate into which his loved little human comrade was about to blunder. Micky himself could have gotten out of the way with no trouble at all. Not only was he almost as quick as the snake, but he was not in the snake's path. He was at one side of it, as he jogged along with his friend. But in another moment the child would step unseeingly on the serpent.

There was not time to shove the little boy out of the way. The snake was about to strike, and nothing is swifter than a striking snake. With a snarl of fury, Micky hurled himself straight at the rattlesnake's head. As the dog sprang, the snake struck. His fangs buried themselves deep in the hero dog's throat. Injected in such a vital spot, the venom was certain to cause death. With a dog's strange knowledge of such things, Micky must have realized that. But he did not hesitate to make the sacrifice.

In another instant, Micky's own fangs were tearing through the snake's body, breaking the squirming back and shaking the ugly creature to pieces. It was too late for him to save his own valiant life, but before he died he made certain that the reptile should not harm the child.

And here is a story of another Boston terrier. A woman brought her pet Boston to a ranch in the west where she was to spend the summer. The men of the ranch laughed at the dog's smallness and his merry, gentle ways. One of them said he would sooner have a doll for a pet than such a dog. Another warned the woman that some fierce squirrel might swallow him.

A week later she was out riding, while the dog scampered along at her side. The horse bolted. She could not stop or swerve him. He made for a cliff-top, whence he and she would have had a fall of nearly ninety feet onto a heap of boulders at the base. The little dog knew nothing of horses or of runaways, but he sensed the terrible danger. He launched himself high in air. His teeth met in the nostrils of the runaway horse, and he swung his own wiry body far to one side. The pain and the sideways yank made the horse veer sharply, almost on the very edge of the precipice, and come to a rearing halt. The woman dropped unhurt to the grass.

After that, the ranchmen made no more fun of the brave dog. Indeed they showed their new admiration for him by feeding him so much that he grew unwieldily fat.

These two stories show that the olden spirit of deathless courage and quick thinking have never been bred out of the Boston terrier, in spite of his recent gentleness.

# The Irish Terrier

SHE was only seven months old when I bought her. I named her Chips. She was born and brought up in an Irish terrier kennel, about four miles from Sunnybank. She was auburn red and was what is known to fanciers as smooth-coated. That means her coat was like a hound's or a pointer's, except for the shaggy eyebrows and "foreface" and beard.

There are two varieties of Irish terriers, you know, just as there are two varieties of fox-terriers: the rough and the smooth. The rough Irish terriers have longer and heavier coats, ranging in color from dark red to light buff. Sometimes roughs and smooths are to be found in the same litter. They are not two distinct strains. Roughs are more favored at dog shows.

As I lifted Chips into the car to take her home, I said to her: "This is your car, now. Watch it! Your car, remember. *Watch* it!"

Of course the puppy did not understand the meaning of every word I spoke. But sometimes a dog's instinct will make it understand the general drift of such a speech. I wondered if Chips were such a dog. Very soon I was to find out.

I held her in my lap on the homeward drive. On the way, we stopped at the postoffice. A man brought out our mail, and opened the car door, but he stepped back much faster than he had advanced. Chips snarled furiously and lunged at him with bared teeth. Mind you, she had not been in that car for more than ten minutes. Yet already she knew it was her duty to guard it against intruders. It was the same, ten minutes after that, when we drove into the grounds here at Sunnybank. As we stopped at the house, my superintendent came over to the car and started to lift out the new puppy. But the new puppy did not see it that way. She flew at him in hot rage, the instant his hand touched the car. Yes, decidedly, Chips was one of the dogs that catch the sense of the order, "Watch!"

At once she made herself at home, quickly learning what persons had a right to be here and what persons were outsiders. The day before her arrival, I had had a low square box made for her. I had lined it with an old shooting coat of mine, and had put it in my study. I pointed this box out to Chips and told her it was her bed. I told her this three times, slowly and distinctly, and I made her get into it. That was enough. Ever since then, as soon as she is let into the house at night, Chips makes a wild rush for that box and jumps into it and curls up to sleep. She growls menacingly if any of the other dogs happen to come near it. It is hers; and she loves it.

I lined the box with my shooting coat instead of any other bedding, because I had worn the coat for years and the puppy thus would associate it with me. Dogs gain more impressions from their scenting powers than in all other ways put together. The scent of her master's coat was to teach Chips not only to track me anywhere, but to recognize my other belongings.

Does that sound silly to you? Any expert dog man will tell you it is true. If ever you get a puppy and plan to keep it indoors at night, line its bed with some long-worn garment of your own. Do this before the puppy has seen the bed. It is one of the first lessons for a puppy.

For the same reason, the first plaything I give to our Sunnybank collie puppies, when they are old enough to romp with anything, is one of my worn-out hiking boots. They chew this merrily, use it in their tugs-of-war, and prefer it to any other toy. Meanwhile, without knowing, they are becoming familiar with their master's footsteps and are learning to trail him by their familiarity with the scent of his old shoe. Try this with your next puppy. It is worth while.

I bought Chips to clear the stables of rats. An Irish terrier is one of the best ratters in the world. Catching and killing rats is one of its chief delights in life. No matter how viciously a cornered rat may sink its teeth into the nose of a true Irish terrier, the dog will not flinch, but will go on killing it.

Chips' chief drawback as a ratter was her habit of bringing slain rats to me and laying them at my feet. Perhaps she thought I would like to eat them, or perhaps she did it to show off her own cleverness and courage in catching them. But it was a nuisance, especially when she could not find me. At such times she would leave the rat on the threshold of my study or lay it tenderly in my study chair.

I have broken her of this habit; but I had to do it carefully. I could not punish her for it, lest she think she was punished for killing the rat. In that case, she might not have killed more of them, for she is an obedient little dog and seldom has needed to be reprimanded twice for the same fault.

I have told about Chips, not because the story was dramatic or exciting or had any special point to it, but because it seems to me that it gives certain Irish terrier traits. For Chips is typical of the whole gay race of Irish terriers.

I have said the rough Irish terrier is preferred in dog shows to the smooth. I don't know why. Before going to a dog show, the rough terrier must be "plucked" or "stripped" by an expert. That means its long and shaggy outer coat is pulled out — painlessly, I have heard — leaving the short and harsh undercoat exposed, and thus showing all the outlines of the figure which the outer coat hides.

The Irish terrier is a game little fighter, but he lacks the aggressive quarrelsomeness of some of the other terrier varieties. He is able and ready to protect himself at all times and to look after his own interests.

If trouble comes his way, he won't dodge it. But he doesn't go out of his way to hunt it.

There is a swagger and an air of self-importance about him, which is one of his charms. He does not slink. He struts. But it is a harmless and mighty amusing form of vanity. Under it all, he is a gentleman. He is capable of deathless love for his master and of a jolly friendliness for all accredited friends.

Chips also has a genius for coaxing. Gently, prettily, she will coax to be allowed to go for a drive or for a walk; to be let out of the house or into the house. She will coax daintily and humbly for pardon, when she knows she has done wrong. Notably, when she has killed a cat.

Like the average Irish terrier, she has an inborn hatred for cats. She knows very well that she has been ordered to leave them alone, but the temptation is too overpowering for her to resist. If she sees a cat when I am with her, she comes back from her dash at the feline the first time I call. But she quivers all over with baffled yearning to go on with the chase. When nobody is around to stop her, she shakes to death every cat she can find. I am told some Irish terriers don't have this impulse, but I have found that most of them do have it—and yield to it.

Nobody knows exactly when and how and where the Irish terrier started his career. Mr. Ridgway, of Ireland, claims there are references to the breed in ancient Irish manuscripts. I can't answer for that. I have never seen these manuscripts, nor copies of them. But many enthusiasts have claimed the Irish was one of the oldest breeds in his own green Island. I doubt if he was found in the long ago in his present trim form.

A red dog, with other likenesses to the Irish terrier besides his mere color, was painted on the cover of a very old Egyptian sarcophagus. Fanciers have taken hold of this doubtful bit of evidence, too, as a part proof that their favorite terrier was loved and honored by kings in the days of the Pharaohs.

A much more probable theory is that he is descended from the old-time Irish wolfhound, and that shrewd breeding dwarfed his size after a few generations. Wiser dog men than I have stated this; and I am very strongly inclined to believe it. I have seen Chips and other Irish terriers in poses where they bear an incredibly strong resemblance to miniature Irish wolfhounds. The color, too, is much the same as the tawny reddish hue ascribed to the original wolfhounds of Erin.

At any rate, nearly sixty years ago the Irish terrier made his appearance in a dog show at Dublin. He was not as compact and graceful and as small, for the most part, as the Irish terrier you see today. But he was handsome enough and interesting enough to catch the fancy of many dog men. They set about to standardize his appearance. Presently he was appearing at dog shows everywhere and making a hit.

The breed was popular in Ireland and then in England; but it was rather a long time in finding its way to America. I have the photograph of the first Irish terrier exhibited in America. That was in 1880. She resembled our modern Irish terriers as a blurred snapshot resembles the beautiful scene it is supposed to depict. Her name was Kathleen; and she was heavy and short-headed and almost tailless. But she was good enough to interest the American fanciers in the breed. More and better Irish terriers were imported into the United States; and more and better Irish terriers were bred here every year. But the original spirit and swagger and sense of fun were never bred out of them. Dr. Bruette sizes up the breed and its early uses in better words than mine:

"The Irish terrier is a worthy product of the country whose name he bears. In Ireland, he is used for hunting foxes and vermin and rabbits. He has no superior as a companion. He is a game all-round sporting proposition, ready to take an active part in anything resembling sport or pleasure. The breed is undoubtedly very old; although there is the usual mystery about its exact origin."

Back in the 1880's, Mr. Ridgway had a female Irish terrier named Antrim Bess. She not only won many prizes on the bench, but had a record for rat-killing. A contest was arranged, to see how many rats Bess could kill in a given number of minutes. I tell the story because it shows an Irish terrier's quick wit and powers of reasoning, not because it describes a slaughter.

The rats were massed in a cage and were let out through a hole at one end. Bess waited outside for the door of the hole to be opened and the rats to be scared out into the open where she could get at them. At a signal the door was opened and the rats started out. Instantly Bess was among them, killing right and left. She allowed none of them to get past her.

But they came out faster than she could catch them with any ease. So the clever little dog thrust her hairy shoulder against the hole, stopping it up. Then, at intervals, she would remove her shoulder from the hole just long enough to let one rat come out. Clamping her shoulder again to the hole, she would kill the fugitive and then move far enough to let the next rat escape.

Mr. Ridgway left no record, that I can find, of the length of time it took Bess to slay all the rats in the trap. But there were eighteen of them, and she killed them all. Not one got away from her, thanks to her nimble wit in thinking of shutting the hole when they came out too fast. I doubt if dogs of most breeds would have had the shrewdness to think of that trick.

In Ireland, too, farmers used Irish terriers to catch foxes which were destroying the chickens and other poultry. Unerringly the terriers used to find where the foxes had their burrows. Down they would dig, excitedly, until the fox was unearthed. Then would follow a spectacular battle in which the fox always was the loser.

This is not as strange as it may seem. For the average fox weighs only eleven or twelve pounds, in spite of his huge coat and brush; while the Irish terrier usually weighs anywhere from twenty-two to twenty-four pounds. The fox may look bigger, but he weighs only about half as much as his adversary. Still, he is a fighter, every ounce of him; and the terrier usually showed signs of the strife.

By the way, the Irish terrier is another dog that can live in moderate comfort and health in a big city, if he be rightly fed and exercised. But he really belongs in the country—as do all dogs that are worthy of the name.

# The Great Dane

LOOK at the accompanying portrait of a Great Dane. It is a perfect likeness, in every detail. It is finely done, but that is not why I ask you to study it carefully and to memorize it. I have another reason.

If some young person happens to see this picture ten or fifteen years from now, he is going to stare at it in amazement. It will not look at all like any Great Dane he has seen. For the Great Dane of ten or fifteen years hence will look entirely different in head and face from the grand dog in this picture.

I will explain what I mean by this change. You will notice that the Dane's ears stand erect above his head in two points. Do you imagine Great Danes are born with ears which look like that, shaped like spear points and standing bolt upright? Well, they are not; and they never have been. Man made those ears. By nature, the Great Dane has big and long and low-flapping ears, very much like the ears of a hound. When he is a puppy, the ears are cut into the shape you see in the picture. Then the ends are fastened upright. While healing, the cut ears are examined every few days. When the edges are found to be puckered together, they are torn loose and made to stand up, in clamps or otherwise, until the edges heal and the ears have learned to grow in their new shape.

Yes, cutting off the greater part of the ear and training the rest of it into shape is cruel and it is painful. But fashion demanded it. So every Great Dane that was going to be shown or owned by an ordinary purchaser was operated on in this torturing way. It went on many years.

As a result, the general public came to suppose that a Great Dane's ears grow in that shape naturally. They do not. Neither do the prick-ears of a Doberman pinscher. Neither do the prick-ears of many Boston terriers; nor of the bull-terriers, nor of certain other breeds whose ears stand pointedly upright. It is all done by hand.

A great many years ago, in England, the Prince of Wales, who was afterward King Edward VII, protested against such unnecessary cruelty. So the British fanciers for the most part stopped cropping dogs' ears. The Great Dane's long and flapping ears were allowed to grow as nature intended.

But here in America the barbarous practice of ear-cropping continued. I wrote many articles denouncing it. So did other humane dog fanciers. But, for a while, that was all the good it did. We kept on. We roused public indignation at last. Bills were introduced in one state after another, to make it illegal to crop the ears of any dog. The ear-croppers fought fiercely against the proposed law. But one state after another made the bill a law. Massachusetts and New York and New Jersey were the first three states to legislate against the cropped ear, and here it became illegal to exhibit a dog whose ears are cropped. More dog shows are held in these states than in any other three, and though the law was enacted only a little while ago, already the effects are to be seen.

Nowadays, when a Great Dane puppy is born, generally his ears are left uncropped, and allowed to hang down, wide and flapping. It gives him an entirely different look and expression, and makes him almost unrecognizable as a Great Dane. He looks more like a big hound than like a Dane. But he is no less beautiful than before, and has escaped much torture. It gives him a much friendlier and more gentle appearance, as nature intended him to have — much as the fierce expression of the former Kaiser of Germany would be lessened if his bristling mustache should be shaved off.

But, whatever is done to his ears, the Great Dane is one of the grandest of dogs. He lacks the slow heaviness of the St. Bernard. He is wise; he has a keen sense of fun; and, rightly treated, he is friendly and gentle. Of course, he is a savage and diligent watchdog, but he shows no savagery to those he likes.

In old dog books — notably by the naturalist, Buffon — the Great Dane is described by the name *le Grand Dancis;* while the Dalmatian (the spotted hound that once was called the coach dog) is described as *le Petit Dancis*. This is French for "the Great Dane" and "the Little Dane." When you see the Great Dane with pendant ears, you can understand why he was once supposed to resemble the lop-eared Dalmatian, or coach dog, and why the latter was nicknamed the Little Dane. There is a very real likeness between the two; though the coach dog usually is spotted black-and-white, while the Dane is found in several different colors.

The Great Dane is associated much more with Germany than with Denmark; and he was brought to his present perfection chiefly by German breeders. A magnificent job they made of it, these German fanciers; but a still better job has been made of it by the American fanciers who have followed them.

Incidentally, the Great Dane is one of the oldest breeds of dogs known. You will find his pictures carved in Greek and Roman ruins. In these likenesses, he is almost exactly the same in appearance as he is today. In these carvings, too, he is given a place of honor above that of any other breed of ancient dog; and it is shown that he was regarded as the foremost dog in the world.

He was the companion of kings; and one of his breed usually had a place at the foot of the throne. In England, in early days, he was known as the "alaunt." And, to judge from early English drawings and paintings, even then his ears usually were cropped. Some pictures, elsewhere and earlier, show him with drooping ears. During the Middle Ages, monarchs used the Great Dane for hunting, as pictures and poems tell. Often these Danes were valued so highly by their royal owners that they were left at home when such dangerous brutes as wild boar were hunted.

As far back as the earliest clear records, the Dane was colored in much the same way as he is now. There were the harlequins (irregularly marked with black and white like the dog in the accompanying portrait), fawn color, bluish-gray and, occasionally, black. They were the largest dogs of their time, even bigger than the giant mastiffs of old. And they were unconquerable fighters.

Ancient and honorable as was the record of the mighty breed, first in Great Britain and then in Germany, the Great Dane was slow to gain any kind of recognition in America. A very few of them were imported from time to time; and they were stared at as freaks. But no regular recognition of them was made by the American dog show magnates until 1886. One or two had been entered for a show a few years earlier, but they got into a fight and several men were bitten in trying to part them. As a penalty, they were barred from that and other shows.

But the Great Dane was here to stay. He was one of the most magnificent animals ever seen. Steadily he grew in favor. Soon he was seen at every large dog show and in ever-increasing numbers. Soon, too, he was seen on country estates and on the streets. "Bob" Fitzsimmons had one of the first Great Danes here, a jet-black dog named Yarrum. He was at his heels on all his walks in both city and country. That was nearly forty years ago; and as many bystanders turned to gape at the enormous dog as at the champion heavyweight prize fighter who owned him. Fitzsimmons and Yarrum were the most spectacular sights on Broadway at one time. The grand old dog was killed by a stroke of lightning; so Fitzsimmons told me.

In my youth, a druggist on Sixth Avenue in New York City owned a glorious dark gray Great Dane named Jake. The dog used to be standing in front of the drug store when I passed there on my way to and from the office. I used to stop and talk to him and to his master; and we three got to be good friends.

Jake was as gentle as he was fearless. He was loved by everyone in the neighborhood. When charity drives were in progress, his master used to tie a little basket with a placard on it, to Jake's neck; and the dog would pace solemnly up and down the street. That was before the days of motor-cars, so Jake's rambles were safe from disaster. He would return home with the basket full of coins and bills, for few pedestrians could resist his friendly stateliness when he approached them for alms. I don't know how many hundred dollars he collected for charity from time to time, but the amount was so great that all the local newspapers made mention of it.

Jake slept in a little room behind the store, the same room in which the druggist kept his safe. As the man was interested in many charities, the safe often contained a goodly sum of money. This fact became known, and two robbers decided to take advantage of it.

One night, they broke into the store and made their way to the back room where the safe was. But the safe's staunch guard was there, too. Jake met them on the threshold. Gone was his usual friendliness. He was guarding his owner's property. Roaring in anger, he charged the two intruders. They were panic-stricken. There was no time for them to escape, so they drew their pistols and opened fire on the attacking dog. Their bullets riddled his body. But he inflicted terrible punishment on them, before the sound of the shots and of his growling brought the nearest policeman on the run to the drug store.

The thieves were captured, thanks to the Great Dane's strength and courage and loyalty. But Jake had saved his master's treasure at the cost of his own life. He died from the bullet wounds he received during the battle. All the neighborhood mourned him.

Perhaps you have read the tales Tolstoi, the Russian novelist and philosopher, wrote about his black non-pedigreed Great Dane, Bulka. The dog was given to Tolstoi in puppyhood and worshipped the novelist for the rest of his long life. Never willingly would he stir from his owner's side.

The Dane was little more than a puppy when Tolstoi set out on a long journey, traveling by stagecoach. The day was boiling hot. Bulka was locked in an upper room to keep him from following his master, but he saw the coach start, and he leaped out of the window in pursuit. He landed on his head on the lawn with a force that stunned him. As soon as he could get up, he staggered after the coach.

The first part of the journey was thirteen miles. There at a post house the hot and weary horses were changed for fresh ones. As Tolstoi stepped out of the coach, Bulka came rushing up to him. The dog had trailed the horses through the burning heat, and had reached the post house almost as they did. Tolstoi wrote of this exploit:

"He leaped up on me, and then threw himself on the ground in the shade of the coach. His tongue lolled out at full length. He was panting violently. He couldn't get his breath. His sides labored. But, all the time, he pounded the ground with his tail, in ecstatic happiness at being with me once more."

Again, Bulka escaped from home while his master was far out in the Russian forests on a bear hunt. The dogs had just surrounded a bear which they had brought to bay, when Bulka — having broken the chain that bound him to his kennel and having caught his master's trail and followed it — dashed up. He burst through the ring of lesser and more cautious dogs and leaped for the bear's face. Said Tolstoi:

"The bear pounded and ripped at him with its claws; hugged him; shook him from side to side; but could not get rid of the grip Bulka had gained on Bruin's cheek and ear. Then the bear stood on its head and shoulders in an effort to pin Bulka to the earth and crush him. But Bulka held on."

Even when Tolstoi saved the rash Dane's life by shooting the bear, the other hunters had to sluice cold water over Bulka to make him let go his death hold on his enemy.

Another time, Bulka got free from home, where he had been shut up, and followed Tolstoi to a wild-boar hunt. There he almost paid with his life for his eagerness to attack a vicious boar that confronted his master. Bulka was badly gashed and torn. Weeks passed before he was well again.

In his old age, Bulka evidently felt death coming on. He had been ailing for days. One evening at sunset he got up and lurched feebly across to where Tolstoi sat at his desk. The Dane licked his master's hand and stared long and lovingly up into his eyes, then walked tremblingly from the room and from the house. Tolstoi wrote, long afterward, concerning this farewell:

"I traveled over the whole region looking for Bulka and I made inquiries everywhere. Nobody had seen him. I could learn nothing as to where he had gone nor how he had died. Probably he made his way to the thickest part of the forest and lay down there to perish."

Great Danes have been the chums of other famous men besides Tolstoi. Bismarck never went anywhere without his Great Dane close at his side. Several of his portraits are painted with this giant dog. Other celebrities have chosen the Dane for their chum, and regarded him as their closest and best loved friend.

But don't buy a Great Dane unless you have plenty of outdoor space in the country for him to exercise in. I would sooner shoot one of the splendid dogs than to keep him cooped up in a city apartment or house.

# The Sealyham

LONG ago, in Scotland, I used to find trouble in telling the Sealyham from his close relative, the West Highland terrier. I find other people who have the same trouble in remembering which is which.

It was an old Scotch dog man who taught me to get them straight in my memory. His teachings may do the same for you. He said:

"If they've got high ears, they're Highlanders. Remember the two words, *high* and *Highlanders*. They go together. The prick-eared tyke is the West Highlander. The one with the flappy ears is the Sealyham."

Never again did I forget. Never again, I think, will you.

Like nearly every variety of terrier, except the Boston, the Sealyham started life in the British Isles. That is the land of the terrier; the land where perhaps more terriers have been bred and evolved and cherished than in all the rest of the world put together. I don't know just what it is that makes a Britisher love a terrier better than he loves any other dog, but I think he does.

The Sealyham you see at shows or curled up on a friend's couch does not strike you as anything much but an unusually pretty and affectionate and amusing pet. And I am sorry to say that seems to be the chief use of the breed here. Their friendliness and cuddlable appearance are perhaps responsible for it. That and the fact that few people here have any chance to put the Sealyham to his true use, even if they happened to know what his true use is; which most of them don't.

It would have made the old north British Sealyham breeders turn over in their graves to see their favorite tyke debased to dozing on cushions and being fondled and getting his airings on the end of a fancy leash in a crowded street. It was not for such purposes they developed him and took such pains over him.

Of all the working terriers, in early days none outranked the Sealyham for power and for grit and for skill in digging down to a larger foe that had hidden underground and for overcoming him there. Study your own Sealyham more closely and you will be able to comprehend why this was so.

He is built on lines that are heavy and powerful, but without a hint of slowness or awkwardness. He is a "utility" dog, every compact inch of him. A dog to jog all day behind the hound pack, and then to conquer the prey which the bigger dogs at last run to earth. A dog that knows no fear and can stand fatigue.

A man who has spent much of his life in England and Scotland has written a fine description of the Sealyham; and in it he deplores the fact that fashion has taken up the good little dog and done so much to keep him from his rightful heritage in the few years since he became popular in the United States. He writes:

"For many years this attractive class of terriers has been carefully bred by a small group of British sportsmen who have cherished the Sealyham for his admirable working qualities and have never been interested in the fads and mandates of the bench-show world.

"No breed is better fitted to 'go to earth' " (get a fox or otter or other creature out of its underground den) "than the Sealyham. Pound for pound, he represents as much dead game courage and as much determination as any dog that lives."

Now here is the story of a strange friendship I witnessed, years back, between a Sealyham and a horse. I say the friendship existed *between* them; but that is hardly correct. The horse adored the Sealyham. There could be no question about that. But I never saw the dog show any special interest in the horse.

It was at a country place where I was visiting. My host had several horses, but there was one fast trotter which was his special favorite. He drove this horse every day and was proud of the beast's speed. He had won several amateur races with him against the trotting horses of his friends.

My host's Scotch groom owned a Sealyham terrier, Jock. For some reason Jock chose this horse's box stall as his bedroom. When I visited the place, Jock and the horse were inseparable pals. At least, Jock spent his nights in the box stall — often curled to sleep in the manger — and the horse had developed a slavish devotion for him.

Whenever the horse was loose in the pasture and Jock appeared, he would canter over to the dog, and nuzzle his head and back. He would then follow meekly wherever Jock might lead the way. Once when Jock was absent at a dog show, the horse stamped and neighed and fidgeted sleeplessly in his stall all night, and refused to eat until his dear friend returned.

One day when my host and I went driving behind this trotter — it was before the era when flint-hard roadbeds and whizzing motor-cars made horse-driving a peril rather than a delight — Jock happened to be running loose. Unnoticed by us, he trotted along, close to the buggy.

From the outset, we were puzzled at the horse's strange behavior. He kept turning his head, trying to look backward. Then he came to an abrupt halt, and turned his head far back for another look. We followed the direction of his glance. That was when first we saw Jock had come with us. The dog had loitered to touch noses with another dog, and the horse would not stir until his comrade trotted alongside again.

Every time the Sealyham stopped or slowed down, the horse would do the same. How the horse knew — unless by sense of smell — that his chum had dropped far behind the buggy, I don't know. But I do know that neither voice nor whip could make him go forward until Jock rejoined us.

Then midway on the trip, a rabbit bounced up from the roadside grass and scuttled at full speed down the road directly in front of us. Jock gave chase, traveling surprisingly fast on his thick little legs. In an instant he was as far ahead of us as the rabbit was far ahead of him. But only for an instant. Immediately the horse broke into a dead run, trying to catch up with his friend. My host was a skilled horseman, but not all his strength or skill could check the runaway. I happened to think of blowing a dog-whistle I carried in my pocket. At once, though reluctantly, Jock gave up the rabbit chase and turned back toward us.

The moment Jock stopped running, the horse slowed to a walk, whinnying a glad welcome to his returning pal. I lifted the dog into the buggy. He rode thus, proudly, for the rest of the trip. The horse, too, behaved well after that, except that he would try to turn his head every now and then to make certain Jock was still near.

Jock's groom-owner got another position, and took his dog with him. The fine horse grieved bitterly for his lost stall-mate. He lost flesh, lost speed, lost appetite, lost all interest in life. To save him from pining away, his master sent to his former employee, offering him his job again at increased pay. Back came the groom, and Jock with him. My host wrote me that the trotter welcomed Jock back to the stall with almost crazy rapture. From that day, the horse was his old-time healthy and spirited self.

If you have a Sealyham or if you get one, treat him as if he were a man's size dog — which at heart he is — not as if he were a pampered Persian kitten.

# The Wire-Haired Fox-Terrier

IT MUST be more than forty years ago that the fox-terrier suddenly became popular in this country. All at once, everyone owned one of them or had friends who did. For centuries the breed had been a favorite in Great Britain.

The fox-terriers in America in those days were not the wire-hairs which are in such tremendous vogue nowadays. They were the so-called smooths — dogs with short and flat coats, not harsh to the touch and not bristling with long, stiff hair.

And fine little fellows they were, alert, gay, loving, perfect companions for anyone who wanted a dog that was not too large. Their chief fault, then, was the tendency to run away. I don't mean all of them had it. But many did. I used to scan the lists of lost dogs in the advertising columns of the newspapers, and the smooth fox-terriers usually led. It was a love of adventure and travel, I think, as much as dissatisfaction with their owners' homes that made many of them wander off and neglect to come back again.

Long afterward, the wire-haired fox-terrier became fashionable here, supplanting his smooth-haired brother. We Americans used to think it was a new variety, just evolved by the fox-terrier breeders; and spoke of it as a novelty, lamenting that some of its specimens seemed to lack the calm good-nature of the old-time smooth-coated dogs. For once in a while the wire-hair is likely to have a hair-trigger temper, which the smooth very seldom had. He doesn't like to be imposed upon.

We were all mistaken. Because the wire-hair had been in existence long years before the smooth was heard of. The smooth was developed from the wire-hair, in Great Britain; not the wire-hair from the smooth-coated dog, as we had always supposed.

I have said, more than once, that Great Britain is the original home and the cradle of nearly all the terrier breeds. Only after some kind of terrier has proven his fine worth over there, do we Yankees discover him and decide whether or not to take him up. It was so with the fox-terrier, as you shall see.

Centuries ago, a terrier was bred and inbred for the purpose of digging foxes out of their earths after the hounds had driven them thither. It was a rough-coated and plucky little fellow that was chosen and built up for this dangerous job in one section of the "hunting country." Because of his prowess with foxes, he became known as the fox-terrier. Elsewhere other terriers were bred for the same purpose, but the name, fox-terrier, stuck to this special variety, and has continued to stick long after its origin is forgotten.

When he was taken up by Fashion in Great Britain, and when he was kept only as a comrade and as a watchdog, the fox-terrier was bred for a while along lines that deprived him of his wiry coat. In other words, the smooth fox-terrier came into existence. More and more people bought him. He came at last to America, where he was made much of. Then, in much later years, his wire-coat was reinstated.

Up to 1882, in the dog shows these re-bred dogs were called broken-coated terriers. Then someone suggested there were so many varieties of terriers with "broken" coats that this kind be known as wire-hairs. The next year — 1883 — saw the first wire-haired dogs shown in this country. But for several decades they had no chance at popularity against the smooth-coated fox-terriers. The wire-hairs were slow in working their way into public favor, but once established, they were here to stay.

One of the reasons for the breed's great advance in people's affections was the fact that the late King Edward of England owned one of them and that he loved it better than any of his many other dogs. When he lay dying, this terrier, Caesar, cried outside his door. Hearing Caesar's moans, the dying king ordered that the dog be admitted to the sick-chamber and that he be allowed to remain with him to the end.

Some days later, a strange and majestic procession wound its way through the streets of London, to solemn music and between throngs of people who stood with their heads uncovered and bowed. King Edward's body was on its way to burial.

All the potentates of the kingdom and of Europe at large rode or walked in that procession, doing final honors to a sovereign they loved and mourned. But ahead of all these potentates and close behind the royal coffin, trotted a little white fox-terrier. Caesar was following his worshipped master to the tomb. Every photograph and painting of that historic scene depicts gallant little Caesar led behind the hearse; unutterably sad, yet showing by his bearing that he had a full sense of his own importance. It is the first time in modern history that any dog has occupied a post of honor, ahead of kings and princes and nobles, at a royal ceremony. For many years afterward, the staunch little dog lived on, loved by all.

I wonder if you have noticed how often fox-terriers are used in animal acts in vaudeville shows? And have you noticed how clever and supple and dashing they are, in these performances? I think they are the best of all canine performers. They learn faster than most others; and they put more spirit into their work. In spite of that, I make it a rule to get up and leave a theater every time an animal act begins. Too often the dogs have been taught their tricks through cruelty.

One of the most famous fox-terriers of recent years was Whitey, whose fame was blazoned in the newspapers, off and on, for a long while. Nobody knows where Whitey came from or who his earlier owner may have been. He limped into the South Division police station, at Buffalo, New York, late one cold night, and across to the desk sergeant.

Ingratiatingly he held up one of his forepaws for the sergeant to look at. It was terribly injured, but the terrier gave no sign of his pain. Hastings, the sergeant, understood and liked dogs. He bandaged and washed the hurt paw, and gave the starving little wire-hair a big dinner and a soft bed near the stove. When the dog was well again, he refused to leave his benefactor or to quit the station house.

The station patrolmen named the waif Whitey and to amuse themselves began to teach him tricks. Whitey proved to be one of those rare dogs that can learn almost faster than they can be taught, and that take a keen pleasure in every trick they can possibly acquire. Presently he was so good at that kind of thing that the tidings of his prowess spread throughout the neighborhood. The newspapers got hold of the story. Whitey was a celebrity.

For the benefit of crowds that came to watch him, he would kneel as if in prayer; play dead dog; open and shut doors at command; fetch coats and hats from their pegs, never once making a mistake in giving them to their rightful owners; and do a dozen other and more complicated stunts. He seemed to revel in the applause and patting he got.

Then one morning Whitey disappeared. The patrolmen and Hastings were miserable. They remembered hearing that fox-terriers sometimes stray from home, but they could not believe their chum had deserted them of his own accord. And he had not.

As he had strolled along the street that day on his morning walk, a dog-catcher had nabbed him. He was taken to the pound. As he did not seem especially valuable, he was condemned to die unless someone should claim him within the next forty-eight hours. Millions of other friendly dogs have had the same treatment.

The police were making inquiries everywhere. So were the newspapers. Someone was found who had seen the dog-catcher seize Whitey. Instantly two patrolmen started at a run for the dog-pound. There among a throng of shivering and heartsick dogs, they discovered their beloved Whitey.

When a crying boy comes to a dog-pound to get his lost pet, he may or may not receive civil treatment from the employees in charge. But when two uniformed patrolmen, with fierce determination blazing in their eyes, make the same demand, it is quite a different matter. Whitey was turned over to the police as soon as they announced that they had come to take him back to the station. Whitey's fine was paid. A license was bought for him. So was an elaborate collar.

Much as the dog liked all his friends at the station house, Hastings was his real master. The wire-hair never forgot it was Hastings who had fed and healed him that cold night when he had limped into the station. Always he rode by Hastings' side in his car. Always he made himself comfortable in Hastings' home, yet he could not be made to stay for any length of time. Almost as soon as he arrived there, a sense of duty sent him racing back on foot to the station. There he would jump up on Hastings' desk and mount guard.

He paid no heed to the patrolmen who went in and out. But no civilian could leave the room without Whitey raising such a din as to bring policemen in from the rear part of the station house to see what was the matter.

More than once when prisoners were brought in, fighting drunk or otherwise disorderly, Whitey would fly at them and aid his police friends to subdue them by inflicting a series of painful bites. Yes, Whitey took his duties as police-assistant very seriously.

The last I heard of Whitey was in an advertisement of a vaudeville show. The advertisement billed him as *The Canine Star.* How he chanced to leave the police station I don't know, for earlier some man had offered the patrolmen fifty dollars for him, which had been declined unanimously.

I have told you that the fox-terrier is centuries old, but the first of the breed to win nation-wide fame was Pitch, a wonderful specimen owned by one Colonel Thornton of England, from 1785 to 1790. Nowadays nobody knows who Pitch's ancestors were, or anything about his pedigree, but his descendants are everywhere. Most of the best fox-terriers, smooth and wire-haired alike, are descended from him. I saw a portrait of him, painted in 1797 by one Gilpin. He would not get a prize today in any orthodox dog show, perhaps. But probably Adam would not win a prize in a fashion show if he were alive now.

In any case, Pitch was a true fox-terrier, and the founder of a splendid modern race. Peace to his memory!

# The Cairn Terrier

AT FIRST glance he looks a good deal like the Scottie. But look again and you will find that he is somewhat smaller, that his coat as a rule is lighter in color and that there are many minor differences in his head and face and build.

Do you know how he got his name? By this time you may be tired of having me remind you that nearly all terriers were used originally to dig vermin and predatory wild animals out of their dens underground. When the prey crawled into crevices in stone-piles, it was not so easy for the larger terriers to follow. Heaps of stones cannot be dug away like soft earth; and often the animal seeking refuge in them was so much smaller than the pursuing terrier that the dog was baffled.

So sportsmen began to breed a terrier which would be small enough to wriggle its way into any rock crevice a fox could crawl through; and which at the same time would have the gameness and the strength to battle with the foe after finding him. This was a hard order to fill. It was easy to breed a small enough dog. It was quite another matter to breed a small dog with all the needed strength and courage. But the sportsmen kept on. At last they were successful.

The rock-heaps in British meadows were often known as cairns. Thus the new dog was called the cairn terrier from his ability to dive into these stone-piles and to work his way through them. He was too short-legged to keep up with the hunt when it was going at any speed. So he was carried in a basket or over the saddle-bow until the prey sought refuge in a cairn or a loose stone wall. Then he was set on the ground and his part in the hunt began.

Unflinchingly, he would catch and follow the scent of the creature which had crawled in among the rocks. His light, powerful, and supple body could carry him far through the loose stones. His punishing jaws and his genius as a fighter could be counted on to do the rest. Like the Scottie, the cairn would emerge with his dead enemy, or he would not emerge at all.

It was long before the cairn was regarded as anything except a mighty useful dog for hunters. He was not a pet. Above all, he was not a lap dog or a favorite of fashion. But he was too good a comrade and house dog to remain forever in the corner of some sporting kennel. At last the fanciers discovered him. Cairn Terrier Clubs were formed in Europe; and the breed was developed along lines laid down by fashion. No longer was the good little dog taken out at dawn to the hunting field. No longer must he risk his life in a dark twisting tunnel of stones where an enemy crouched waiting for him. His ancient job was gone.

But the spirit and the brain that had fitted him for that hazardous job were not gone, no matter what changes the fanciers might make in his outward looks. The cairn remained the same pocket-edition of a warrior and strategist he always had been. Never had he been savage or mean-tempered. Always he had been a jolly and friendly dog at heart. In short, he was an ideal companion for anyone.

Cairns were soon brought to the United States. They won almost immediate favor. Clubs were formed for their advancement. Every sizable dog show had a long benchful of them. The cairn had come to America to stay. He was very much at home here; and every year more people bought or bred him.

I think it is the cairn's queer originality which has done more than anything else to win him his way, outside the hunting field. He does his own thinking; and sometimes he does it along decidedly strange lines. Let me tell you one true story to show what I mean.

His name was Jerry. He lived in Belfast, Ireland, where he was kept as a house dog. But as time went on, the household saw less and less of him. For Jerry very seldom was at home. He had discovered a most ghastly form of amusement, and it took most of his spare time. It won for him much newspaper notoriety; and it won for him throughout the whole region the nickname of "the funeral dog."

I don't know how or when Jerry gained his taste for attending funerals, but it became a craze with him. For two or three years he was to be seen at almost every funeral held within many miles of Belfast. He seemed to know by instinct just when there was to be a funeral and just where it was going to be. Merrily he would join the procession and follow to the cemetery. Throughout the services, he would stand close beside the grave, wagging his tail and grinning up at the mourners. Sometimes he would be lifted by the scruff of the neck and carried away. But inside of a minute or two, he was back again.

Watching him at these sad ceremonies, nobody could doubt the little cairn was having a jolly time. Nobody has ever been able to guess why. In all other ways Jerry was a normal house dog. But funerals were as much of a spree to him as is cat-chasing to the Irish terrier.

He learned to associate undertakers with funerals. Though he had formerly been a one-man dog, yet now he made friends eagerly with every undertaker whose shop he could enter. He found out, in course of time, where at least twenty undertakers lived or worked. He would make the rounds, visiting all of them and coaxing them to pet him or to take him for walks.

The grave-diggers of several parishes learned to know Jerry by sight, and to look forward to his presence at funerals. They made a pet of him, and fed him so much and so often from their dinner pails, that he had no need to go back home for food. His owner was disgusted at the cairn's queerness, and refused to keep him longer.

But disinheritance did not bother Jerry at all. If his master had cast him off, he had plenty of other friends, enough to eat, and much to amuse him. He was a vain little fellow; and he grew to be vastly flattered by the attention he attracted at burials. That much of this attention was unfavorable — not to say hostile — annoyed him no more than did his master's disowning him.

Once he traveled thirty miles, from Belfast to Donaghadee, in the wake of a funeral procession; and he followed another funeral ten miles, from Belfast to Lisburn. These were but two of the scores of long and hard journeys his thick little legs carried him on, just for the pleasure of acting as reception committee at graves.

But his funeral career came to a sudden end. Numbers of people hotly resented the laughable presence of Jerry at the funerals of their loved ones. They complained to the authorities, demanding that the cairn be prevented from behaving in this scandalous fashion. The complaints became so many and so vehement that the Mayor and Council of Belfast were forced to take action in the matter.

Jerry was captured at the next funeral he attended and taken to the dog-pound. There he was condemned to death. He seemed doomed to pay with his life for his crazy ideas of a good time.

But the Irish have a rich sense of fun. Jerry's exploits had brought local newspaper readers many a good laugh. When word of Jerry's death sentence was printed in these papers, there was a storm of protest. All at once Jerry became a Public Issue. True, the complaints against him were well justified; but he had won for himself a host of fun-loving admirers.

The upshot of it all was that Jerry was set free from the pound, on condition that he be not allowed to attend any more funerals. So his friends bought him a life-lodging at the Belfast Dogs' Home. There he lived in comfort and even in luxury, much petted and often visited, until his death some years later.

But most cairns find a less eccentric way of enjoying life and of planning their own daily routine than did Jerry. In addition to their wit and originality, they have a keen loyalty to the master they serve. This loyalty keeps them at his side and makes them guard his home and his possessions, even while their sense of humor makes them the best of companions.

Whether in a crowded city or nosing at rabbits in the cracks of a farm's stone wall, the cairn is self-reliant and at home. In fact, he is at home under nearly all conditions, as long as he is not cuddled and pampered like a lap dog.

# The Airedale

I CAN remember well when there was no such dog as the Airedale terrier. There was not a single Airedale on earth. Perhaps you would like to know how he came into existence? It is rather interesting, I think.

I have said there were no such dogs as Airedales. But there were bulldogs, bull-terriers, and pit bull-terriers and otter-hounds; and there were mixed-breed dogs that were terrors at fighting and at helping poachers steal game and at slaying vermin. From this formidable bunch of dogs, the Airedale was developed.

In Yorkshire, England, and in southern Scotland and in other parts of Great Britain where mill-hands and miners and quarrymen labored, dog-fighting was the favorite pastime. Along the Aire River, the cruel sport was at its zenith. On Sundays groups of men from one factory or mine or mill would take their best dog over to some group from another industrial center to pit him against their most warlike dog. This went on for many years.

It was a matter of local pride to have the best fighting dog in the region. So breeders figured out the strongest fighting qualities in various kinds of dogs, and began to combine these qualities by careful cross-breeding. Sometimes the experiment was a failure. But gradually they were able to find what they sought — a breed combining the grit and pugnaciousness and strength and skill and endurance of the several breeds among its ancestors.

At last, among the factory villages along the Aire, a dog was produced which was the finest fighting machine for his size at that time. Moreover, thanks to his strain of otter-hound blood, he was good for hunting. That is a rare combination. Many a shilling did his breeders win on him in the Sunday inter-factory fights. Many a pheasant or rabbit did his lawless breeders kill, on game preserves, through his skill as a hunter. At last the Aire men had found the ideal dog for their use.

His fame spread wherever brutal dog-fighting or poaching were favorite sports. He was bred in ever-increasing numbers. From his birthplace, he took the name Airedale. But you would have to look at him twice, and then fifty times more, to recognize him as the stylishly handsome and upstanding Airedale of today.

He weighed forty pounds, at that time, and every ounce of the forty pounds was a vibrant mass of fighting energy. He, and his ancestors for several generations, had been bred to fight at the drop of a hat — or without waiting for such a signal. He was a murderously efficient battler and a crafty hunter — the kind of dog which rejoiced the toughs who bred and owned and used him.

Word came to London that a new forty-pound terrier had been developed. Fanciers laughed at this news, declaring no terrier could weigh that much (though nowadays it is far less than the standard Airedale weight), and saying the new dog was more probably a hound. This statement was partly correct, for there was hound blood in the Airedale, much more then than there is now — sturdy, fierce otter-hound blood.

But his fame was spreading. A new set of breeders took him up; men who tried to make his body symmetrical and to give him a brain which could master other things besides fighting. The task was easy, for the dog was clever and adaptable. In the course of time, his murderous temper softened into fine spirit. The original breeders stared in amazement to find Airedales the pets of women and children; and shook their heads at the tremendous change which had come over the terrible battlers in the course of a few generations. But the British fanciers had taken up the Airedale, and with every passing generation he was a better dog.

As far as I can find out, the first Airedale to be brought to America for show purposes— perhaps the first Airedale to be brought here at all — was one Bruce. Mason imported him to New York in 1891. The dog did not make any kind of a hit with the Yankee fanciers. Soon afterward he was sold at a dog auction for twenty-one dollars.

For years after Bruce's debut, very few Airedales were seen here. Not until 1898 were they recognized as a distinct breed at dog shows. But from that time, there was no stopping the newcomers. All classes of people took them up.

Though he is kept now chiefly as a pal and a guard and for bench-show purposes, the present-day Airedale has lost none of his rowdy ancestors' prowess as a hunter. Of late, the few owners who have taken the trouble to "field break" him, report that he is well up to the average of most well-trained bird dogs, in every way. Especially is he good at duck-hunting. Here is where his otter-hound strain stands him in good stead. It makes him a fine swimmer and retriever. In all weathers he will plunge into the water—even if the water be choked with floating ice — to bring in a duck that has been killed or wounded. Hardy and brave and wise, he is of great value to the sportsman who takes the trouble to train him to hunt.

The Airedale has been sneered at by his detractors as a "standardized mongrel." If that is true, then the Boston terrier is also a standardized mongrel. So is every breed of dog that does not date back unbrokenly in his present form to the beginning of the world. So, most of all, are we humans. Few of us can claim unbroken descent for many generations from any one nationality or type.

So let us admit that the Airedale is a member of a distinct breed—

a grand breed, at that — and not a "standardized" anything. But why the term mongrel should be used as a reproach, I never have been able to understand. Some of the wisest and greatest dogs of my acquaintance have been mongrels. They have been my best chums. And I am not by any means the only authority who says so. Among those who agree with me is the celebrated Dr. G. W. Little, an acknowledged dog expert. He writes:

"The mongrel is often the superior of the well-bred dog in intelligence; and is less subject to disease."

Of course, there are mongrels and mongrels. But if you own one of the good mongrels — and most of them are good if they get half a chance — you own a splendid dog. Do not say: "Oh, he's just a mutt, of course, but we're fond of him." Don't be ashamed of him. But be ashamed of yourself if you have not brought out all the great qualities which lie dormant in his clever brain and hardy body.

Among the dogs which did best and most heroic service during the World War were the mongrels — crossbreeds is a more correct and kinder term for them. Many of the best pal dogs and farm dogs and trained dogs on record have been crossbreeds. There are more crossbred dogs everywhere in the world, than all thoroughbred dogs put together. This is a digression, but it is worth remembering next time you hear your crossbred dog called a mutt.

Now let us get back to the Airedale, shan't we? I want to tell you a story, if I may, which brings out some of his unusual traits, and some of the thought and action of which he is so capable if he be given half a chance.

He was a big Airedale, named Kentucky Boy. They called him Boy, for short. He belonged to Robert M. Byrne, a Los Angeles man. One summer afternoon, Byrne took Boy down Hollywood Boulevard for a walk. In obedience to police rules, he had the dog on a leash. Suddenly Boy strained at the leash, whimpering and trembling. Before Byrne could tighten his hold, Boy had yanked the leash out of his master's hand and was tearing down the street at top speed, barking loudly. For the first time in his life, he refused to come back when Byrne called him. This was such a strange way for the well-trained Airedale to behave that Byrne ran after him to see what might be the reason for his unwonted excitement.

At the front door of a picture studio building, Boy halted. Throwing himself against the door, he barked and screamed, scratching and biting the panels. A curious crowd gathered. Again Byrne called his dog back. The dog paid no attention to anything except his frantic assault on the studio door. A policeman hurried up.

Just then a snakelike wisp of smoke curled out from the crack at the bottom of the door. It was the very first sign that fire had broken out inside the building. How had Kentucky Boy, two blocks away, guessed there was fire and a need to give the alarm? That is one of the million canine puzzles nobody knows how to solve.

Flames were roaring in one wing of the building. In another five minutes the fire would have gained such headway that nothing could have stopped it. Thanks to Boy, the fire department was summoned in time to keep the place from being a total wreck and to keep the blaze from spreading to a packed motion picture theater next door.

Boy's exploit was rewarded by the State Commission for the Protection of Children and Animals. He received from the Commission a big silver medal. He received from his master something he valued more than fifty medals — a quart of ice cream!

By the way, a few weeks earlier, Mr. Byrne had been at work on his lawn when Boy rushed out of the house and flung himself madly against his master, thrusting him to one side — barely a second before a gigantic dead tree on the lawn crashed down upon the spot where Byrne had been at work.

I could write for hours, telling other true stories of glorious deeds of bravery performed by Airedales, but they would only substantiate what I think I have already pointed out concerning the brain-power, prowess, and comradeship of these wonderful dogs.